AF869851

Springer Theses

Recognizing Outstanding Ph.D. Research

Aims and Scope

The series "Springer Theses" brings together a selection of the very best Ph.D. theses from around the world and across the physical sciences. Nominated and endorsed by two recognized specialists, each published volume has been selected for its scientific excellence and the high impact of its contents for the pertinent field of research. For greater accessibility to non-specialists, the published versions include an extended introduction, as well as a foreword by the student's supervisor explaining the special relevance of the work for the field. As a whole, the series will provide a valuable resource both for newcomers to the research fields described, and for other scientists seeking detailed background information on special questions. Finally, it provides an accredited documentation of the valuable contributions made by today's younger generation of scientists.

Theses are accepted into the series by invited nomination only and must fulfill all of the following criteria

- They must be written in good English.
- The topic should fall within the confines of Chemistry, Physics, Earth Sciences, Engineering and related interdisciplinary fields such as Materials, Nanoscience, Chemical Engineering, Complex Systems and Biophysics.
- The work reported in the thesis must represent a significant scientific advance.
- If the thesis includes previously published material, permission to reproduce this must be gained from the respective copyright holder.
- They must have been examined and passed during the 12 months prior to nomination.
- Each thesis should include a foreword by the supervisor outlining the significance of its content.
- The theses should have a clearly defined structure including an introduction accessible to scientists not expert in that particular field.

More information about this series at http://www.springer.com/series/8790

Felix Wolf

Analysis and Implementation of Isogeometric Boundary Elements for Electromagnetism

Doctoral Thesis accepted by
Technische Universität Darmstadt, Germany

Springer

Author
Dr. Felix Wolf
Institute TEMF and Centre
for Computational Engineering,
Computational Electromagnetics Technische
Technical University of Darmstadt
Darmstadt, Germany

Supervisors
Prof. -Ing. Stefan Kurz
Electromagnetism and Mathematical
Modelling
Technische Universität Darmstadt
Darmstadt, Germany

Prof. rer. nat. Sebastian Schöps
Computational Electromagnetics
Technische Universität Darmstadt
Darmstadt, Germany

ISSN 2190-5053 ISSN 2190-5061 (electronic)
Springer Theses
ISBN 978-3-030-61938-1 ISBN 978-3-030-61939-8 (eBook)
https://doi.org/10.1007/978-3-030-61939-8

© The Editor(s) (if applicable) and The Author(s), under exclusive license to Springer Nature Switzerland AG 2021
This work is subject to copyright. All rights are solely and exclusively licensed by the Publisher, whether the whole or part of the material is concerned, specifically the rights of translation, reprinting, reuse of illustrations, recitation, broadcasting, reproduction on microfilms or in any other physical way, and transmission or information storage and retrieval, electronic adaptation, computer software, or by similar or dissimilar methodology now known or hereafter developed.
The use of general descriptive names, registered names, trademarks, service marks, etc. in this publication does not imply, even in the absence of a specific statement, that such names are exempt from the relevant protective laws and regulations and therefore free for general use.
The publisher, the authors and the editors are safe to assume that the advice and information in this book are believed to be true and accurate at the date of publication. Neither the publisher nor the authors or the editors give a warranty, expressed or implied, with respect to the material contained herein or for any errors or omissions that may have been made. The publisher remains neutral with regard to jurisdictional claims in published maps and institutional affiliations.

This Springer imprint is published by the registered company Springer Nature Switzerland AG
The registered company address is: Gewerbestrasse 11, 6330 Cham, Switzerland

It is of no use whatsoever.

Heinrich Hertz,
about possible applications of
electromagnetic waves, cf. [84, p. 83].

Supervisor's Foreword

It is my great pleasure to introduce Dr. Felix Wolf's thesis research, which has been accepted for publication in Springer Theses. Dr. Wolf joined my research group for Electromagnetism and Mathematical Modelling after successfully completing the Master program in Mathematics at Technische Universität Darmstadt in August 2016. He then began his doctoral studies, supported by a scholarship from the German Excellence Initiative. Later he joined the project, "Simulation of Superconducting Cavities by Isogeometric Boundary Elements", funded by the German Research Foundation. He completed his doctoral studies with an oral defense on 19th December 2019. Dr. Wolf received the Young Scientist Award as well as a Best Paper Award in 2018, both from U.R.S.I., the International Union of Radio Science.

Modeling, simulation, and optimization play a key role for engineering products in the electromagnetic domain. However, according to current research, the proportion of grid generation in total simulation time is over 80%, a rather time consuming process.[1] A more recent development to overcome the challenge is the so-called Isogeometric Analysis (IGA). In contrast to conventional approaches that use piecewise low-order polynomials, the geometry and solutions in IGA are represented by splines of higher polynomial order. Geometries created using the tools of Computer-Aided Design (CAD) can therefore be represented exactly without generating any grids. Dr. Wolf adopted this approach for the Boundary Element Method (BEM). The BEM is well suited for linear partial differential equations with constant coefficients. Only the boundary of the problem domain needs to be taken into account. This is a perfect fit with the boundary representation (B-rep) of solids in CAD systems.

Dr. Wolf has been conducting research on Isogeometric Analysis for Boundary Element Methods (IGA-BEM), in particular for electromagnetic scattering and eigenvalue problems. In addition to the more practical advantages mentioned above, the splines provide better solutions than conventional approaches because the smoothness of the fields can be exploited throughout. Dr. Wolf showed

[1] See e.g. Sandia Labs Report 4647.

impressively high convergence orders (>10) in his numerical experiments. Thanks to the modern matrix compression based on H2 matrices, the presented method works also very efficiently.

The research topic is not only relevant in practice, but also mathematically demanding and interesting, especially with regard to functional analysis. The central theoretical result of Dr. Wolf's thesis work concerns the approximation properties of spline spaces with respect to the trace spaces of the Sobolev space $H(\mathrm{curl})$ for piecewise smooth boundaries. This is a new result and has not been proven before. Using this and other results, Dr. Wolf was able to prove existence, uniqueness, and the convergence order of IGA-BEM solutions. He also confirmed theoretical convergence rates through various numerical experiments.

Dr. Wolf has eventually turned to eigenvalue problems as they occur during the simulation of superconducting cavities for particle accelerators. The boundary element method appears unfavorable at first glance because it yields a non-linear eigenvalue problem. Dr. Wolf circumvents this difficulty in a very elegant way, by leveraging the Contour Integral Method (CIM). This is a relatively recent method from complex analysis. It reduces the non-linear eigenvalue problem to a linear one, while retaining the eigenvalues. The combination of CIM and IGA-BEM was successfully accomplished in Dr. Wolf's thesis work, this being the first time someone has done this. The numerical experiments highlight the high order of convergence, which permits a very accurate computation of the eigenvalues. The method presented by Dr. Wolf hence provides a serious alternative to the multitude of existing domain-oriented methods.

Dr. Wolf's thesis represents a considerable advancement in the field of boundary element methods. He published his results as main author in top journals. The software resulting from his thesis work was made available publicly in the software library BEMBEL (the Boundary Element Method Based Engineering Library), www.bembel.eu. The library was created as part of a collaboration between the University of Darmstadt and the University of Basel. Dr. Wolf was not only involved in the provision of content, but also played a leading role in the coordination of the collaboration.

In summary, Dr. Wolf's work provides a significant and original scientific contribution towards closing the gap between numerical analysis, design, and engineering. He has my highest recommendation.

Darmstadt, Germany
April 2020

Stefan Kurz

Parts of this thesis have been published in the following articles:

A. Buffa, J. Dölz, S. Kurz, S. Schöps, R. Vázquez, and F. Wolf. "Multipatch approximation of the de Rham sequence and its traces in isogeometric analysis". In: *Numerische Mathematik* 144, (2020), 201–236.

J. Dölz, H. Harbrecht, S. Kurz, M. Multerer, S. Schöps, and F. Wolf. "Bembel: The Fast Isogeometric Boundary Element C++ Library for Laplace, Helmholtz, and Electric Wave Equation". In: *SoftwareX* 11, (2020), 10476.

J. Dölz, H. Harbrecht, S. Kurz, S. Schöps, and F. Wolf. "A fast isogeometric BEM for the three dimensional Laplace- and Helmholtz problems". In: *Computer Methods in Applied Mechanics and Engineering* 330. Supplement C, (2018), 83–101.

J. Dölz, S. Kurz, S. Schöps, and F. Wolf. "A Numerical Comparison of an Isogeo–metric and a Parametric Higher Order Raviart-Thomas Approach to the Electric Field Integral Equation". In: *IEEE Transactions on Antennas and Propagation* 68.1, (2019), 593–597.

J. Dölz, S. Kurz, S. Schöps, and F. Wolf. "Isogeometric Boundary Elements in Electromagnetism: Rigorous Analysis, Fast Methods, and Examples". In: *SIAM Journal on Scientific Computing* 41.5, (2019), B983–B1010.

S. Kurz, S. Schöps, and F. Wolf. "Towards a Spectral Method of Moments using Computer Aided Design". In: *Advances in Radio Science* 17, (2019), 59–63.

Acknowledgements

Let me first thank Stefan Kurz and Sebastian Schöps for the great advice during the last years. It has been truly invaluable. Their help and assistance went far beyond what can usually be expected during a Ph.D. In this instance, I would also like to thank Christoph Erath who offered me his help should I ever find myself in need of it.

Without the numerical examples this thesis would be much weaker, so I am exceptionally grateful to Jürgen Dölz, Helmut Harbrecht, and Michael Multerer who made their codebase available to me. Without them the implementation as it exists today would not have been possible. I would also like to thank Gerhard Unger for his help and many reference computations for the debugging of the contour integral method.

With regard to the content of this document, I am grateful for all the constructive input of my coauthors during the creation of the various manuscripts. Also I would like to thank Jürgen Dölz, Gerhard Unger, Rafael Vázques, and Annalisa Buffa for the assistance they offered with my manuscripts and related theoretical questions.

With respect to this document itself, I would like to express gratitude to Arthur Windemuth, Mehdi Elasmi, Jürgen Dölz, Thorben Casper, and Hans-Peter Linke for their proofreading and constructive input.

Lastly, I would like to thank all of my colleagues at the Institute for Accelerator Science and Electromagnetic Fields. The work environment and camaraderie here is beautiful and I could not have wished for a better group of people to spend the last three and a half years with.

This work was supported by DFG Grants SCHO1562/3-1 and KU1553/4-1, and the Excellence Initiative of the German Federal and State Governments and the Graduate School of Computational Engineering at TU Darmstadt.

Darmstadt, Germany
December 2019

Felix Wolf

Contents

Symbols

General

$\lesssim, \gtrsim$	$\leq$ and $\geq$ up to a constant factor independent of variables appearing in terms on either side
$\simeq$	$\lesssim$ and $\gtrsim$
$\mathbb{K}$	either $\mathbb{R}$ or $\mathbb{C}$
h	mesh size
m	level of refinement, alternatively size of a matrix
s, t	index referring to the regularity of a Sobolev space
$\times$	Cartesian product
$\otimes$	tensor product
$\square$	open unit square
u_κ^*	Helmholtz fundamental solution
$u_{\kappa,\boldsymbol{\lambda},\boldsymbol{\lambda}'}^*$	localised fundamental solution
η	parameter for the admissibility condition

Geometric Entities

Ω	Lipschitz domain in $\mathbb{R}^3$
Γ	boundary of Lipschitz domain in $\mathbb{R}^3$
$\boldsymbol{n}$	outward-directed unit normal
$\boldsymbol{F}_j$	C^∞ geometry mapping from $\square$ to Γ_j
$\iota_*(\boldsymbol{F})$	for $* = 0, 1, 2$, pull-backs induced by a given diffeomorphism $\boldsymbol{F}$
$\vartheta(\boldsymbol{x})$	surface measure at a given point $\boldsymbol{x}$ on Γ
N	number of patches, alternatively number of quadrature points

$\Gamma_{i,j,k}, \Gamma_\lambda$	element of the nested refinement sequence i.e., a cluster of the cluster tree
$\boldsymbol{F}_\lambda$	localised geometry mapping
$\mathcal{T}$	cluster tree on Γ

Operators

γ_0	Dirichlet trace operator
$\boldsymbol{\gamma}_0$	tangential trace operator
$\boldsymbol{\gamma}_t$	rotated tangential trace operator
$\gamma_{\boldsymbol{n}}$	normal trace operator
div_Γ	surface divergence
grad_Γ	surface gradient
$\mathbf{curl}_\Gamma$	vector-valued surface curl
curl_Γ	scalar surface curl
$\tilde{\Pi}_{p,\Xi}$	one-dimensional multipatch quasi-interpolation operator
$\tilde{\Pi}^\partial_{p,\Xi}$	commuting one-dimensional multipatch quasi-interpolation operator
$\tilde{\Pi}^0_{p,\Xi}$	$H^1(\square)$ conforming, commuting multipatch quasi-interpolation operator for the reference domain
$\tilde{\Pi}^1_{\boldsymbol{p},\Xi}$	$\boldsymbol{H}(\mathrm{div}, \square)$ conforming, commuting multipatch quasi-interpolation operator for the reference domain
$\tilde{\Pi}^2_{p,\Xi}$	$L^2(\square)$ conforming, commuting multipatch quasi-interpolation operator for the reference domain
$\tilde{\Pi}^0_\Gamma$	$H^1(\Gamma)$ conforming, commuting multipatch quasi-interpolation operator for the physical domain
$\tilde{\Pi}^1_\Gamma$	$\boldsymbol{H}(\mathrm{div}, \Gamma)$ conforming, commuting multipatch quasi-interpolation operator for the physical domain
$\tilde{\Pi}^2_\Gamma$	$L^2(\Gamma)$ conforming, commuting multipatch quasi-interpolation operator for the physical domain
π, π_j	quasi-optimal $\boldsymbol{H}(\mathrm{div}_\Gamma, \Gamma)$ projection, defined patchwise via π_j
$\mathcal{P}_s$	orthogonal projection with respect to the $H^s(\Gamma)$ scalar product
$\mathcal{P}_{\mathrm{div}}$	orthogonal projection with respect to the $\boldsymbol{H}^0(\mathrm{div}_\Gamma, \Gamma)$ scalar product
$\mathcal{P}_\times$	orthogonal projection with respect to the $\boldsymbol{H}_\times^{-1/2}(\mathrm{div}_\Gamma, \Gamma)$ scalar product
R	regularising projection
R_0	regularising projection of higher regularity
$\boldsymbol{\mathcal{W}}_\kappa$	Maxwell double layer potential
$\boldsymbol{\mathcal{V}}_\kappa$	Maxwell single layer potential
$\mathbf{V}_\kappa$	Maxwell single layer operator
$\mathbf{V}_{\kappa,h}$	discretised Maxwell single layer operator
$\mathbf{V}^*_{\kappa,h}$	local element matrix

Physical Entities

κ	wavenumber
μ	magnetic permeability
ε	electric permittivity
$\mathcal{E}$	electric field intensity
$\mathcal{H}$	magnetic field intensity
$\mathcal{C}$	electric displacement
$\mathcal{B}$	magnetic induction
$\mathcal{J}$	current density
ϱ	charge density
$\boldsymbol{E}$	time-harmonic electric field intensity
$\boldsymbol{H}$	time-harmonic magnetic field intensity
$\boldsymbol{D}$	time-harmonic electric displacement
$\boldsymbol{B}$	time-harmonic magnetic induction
$\boldsymbol{J}$	time-harmonic current density
$\widehat{\varrho}$	time-harmonic charge density

Sobolev Spaces

$H^s_{\mathrm{pw}}(\Gamma)$	space of functions patchwise in $H^s(\Gamma_j)$
$\boldsymbol{H}^s_{\mathrm{pw}}(\mathrm{div}_\Gamma,\Gamma)$	space of functions patchwise in $\boldsymbol{H}^s(\mathrm{div}_\Gamma,\Gamma_j)$
K^s_j	kernel of $\gamma_{\boldsymbol{n},j}$ w.r.t. $\boldsymbol{H}^s(\mathrm{div},\Gamma_j)$
$(K^s_j)'$	dual of K^s_j w.r.t. $\boldsymbol{H}^0(\mathrm{div}_\Gamma,\Gamma)$
X'	dual space of X
$\langle\cdot,\cdot\rangle_X$	scalar product of the space X
$\langle\cdot,\cdot\rangle_{X\times X'}$	duality pairing between X and X'
$L^2(\Omega)$	space of square-integrable functions on a Lipschitz domain Ω
$H^s(\Omega)$	Sobolev space of regularity s on a Lipschitz domain Ω, for $s=0$ this coincides with $L^2(\Omega)$ and for $s<0$ it denotes the dual w.r.t. $L^2(\Omega)$.
$\boldsymbol{L}^2(\Omega)$	space of componentwise square-integrable, vector-valued functions on a Lipschitz domain Ω
$\boldsymbol{H}^s(\Omega)$	Sobolev space of vector-valued functions of componentwise regularity s on a Lipschitz domain Ω, for $s=0$ this coincides with $L^2(\Omega)$
$\boldsymbol{H}^s(\mathbf{curl},\Omega)$	functions of $\boldsymbol{H}^s(\Omega)$ with curl in $\boldsymbol{H}^s(\Omega)$
$\boldsymbol{H}^s(\mathrm{div},\Omega)$	functions of $\boldsymbol{H}^s(\Omega)$ with divergence in $H^s(\Omega)$
$\|\|\cdot\|\|_X$	norm of the space X
$L^2(\Gamma)$	space of square-integrable functions on a surface Γ

$H^s(\Gamma)$	Sobolev space of regularity s on a surface Γ, for $s=0$ this coincides with $L^2(\Gamma)$ and for $s<0$ it denotes the dual w.r.t. $L^2(\Gamma)$
$\boldsymbol{H}^s(\mathrm{div}_\Gamma,\Gamma)$	functions of $\boldsymbol{H}^s(\Gamma)$ with surface-divergence in $H^s(\Gamma)$
$\boldsymbol{H}_\times^{-1/2}(\mathrm{div}_\Gamma,\Gamma)$	rotated tangential trace space of $\boldsymbol{H}(\mathbf{curl},\Omega)$
$\boldsymbol{H}_\times^s(\Gamma)$	rotated tangential trace of $\boldsymbol{H}^{s+1/2}(\Omega)$ for $0<s<1$, and for $1<s<0$ dual of $\boldsymbol{H}_\times^{-s}(\Gamma)$ w.r.t. $\langle\cdot,\cdot\rangle_\times$
$\langle\cdot,\cdot\rangle_\times$	a special duality pairing

Discrete Spaces

p	polynomial degree
$\boldsymbol{p}$	N-tuples of pairs of polynomial degrees
Ξ	knot vector
Ξ_m	equidistant knot vector with 2^m elements
$\boldsymbol{\Xi}$	N-tuples of pairs of knot vectors
b_i^p	B-spline basis function
$S_p(\Xi)$	B-spline space of degree p over knot vector Ξ
$S_{p_1,p_2}(\Xi_1,\Xi_2)$	tensor product B-spline space of degrees p_1,p_2 over knot vectors Ξ_1,Ξ_2
$\tilde{I},\tilde{Q}$	support extension of I,Q
$\mathbb{S}^0_{\boldsymbol{p},\boldsymbol{\Xi}}(\Gamma)$	$H^1(\Gamma)$-conforming spline space in the physical domain on a multipatch geometry Γ
$\mathbb{S}^1_{\boldsymbol{p},\boldsymbol{\Xi}}(\Gamma)$	$\boldsymbol{H}(\mathrm{div}_\Gamma,\Gamma)$-conforming spline space in the physical domain on a multipatch geometry Γ
$\mathbb{S}^2_{\boldsymbol{p},\boldsymbol{\Xi}}(\Gamma)$	$L^2(\Gamma)$-conforming spline space in the physical domain on a multipatch geometry Γ
$\boldsymbol{Q}_p$	space of quadrilateral Raviart-Thomas elements
$K_j^{\mathbb{S}}$	kernel of $\gamma_{\boldsymbol{n},j}$ w.r.t. $\mathbb{S}^1_{\boldsymbol{p},\boldsymbol{\Xi}}(\Gamma)$
$\mathbb{S}^*_{p,m}(\Gamma)$	globally discontinuous, locally polynomial basis

Chapter 1
Motivation: The Cavity Problem

While many numerical methods for electromagnetism have been understood and established, certain engineering applications are pushing these schemes to their limits.

One such example is the design and simulation of resonator cavities for particle accelerators, for example, the TESLA cavity [1].

These cavities are crucial components of particle accelerators. Within them, resonating electric fields are induced, see Fig. 1.1, such that so-called particle bunches become accelerated to relativistic speeds. The idea behind this principle is explained in Fig. 1.1.

The resonant frequency has to be within a specific range, attuned to the target speed of the particle bunches. This desired operating frequency of these devices is within the gigahertz regime, while changes after production through so-called *tuning* can only be made within a range of 300 kHz [1, Table II].

This tuning mechanism is required to compensate manufacturing errors, as well as errors due to deformations induced by the operation of the device. These effects are known as *Lorentz detuning* and *microphonics* and have an impact of 600 Hz on the Tesla cavity [1, Table II]. Thus, it is desirable that an initial design should fit the desired specifications as closely as possible, such that the tuning range can be reserved exclusively to correct the above mentioned factors and does not have to compensate for errors in the initial design.

While the tuners still can compensate for manufacturing errors and deformations due to Lorentz detuning, conventional numerical methods struggle to resolve changes of this magnitude, since they happen within a relative margin of roughly 10^{-7}, and there are other examples of applications, where such high demands of accuracy are desired of simulations. One other example is presented by Georg et al. [3], who explain that one must resolve eccentric deformations of cavities within a relative error margin of 10^{-6} w.r.t. the operating frequency.

© The Author(s), under exclusive license to Springer Nature Switzerland AG 2021

F. Wolf, *Analysis and Implementation of Isogeometric Boundary Elements for Electromagnetism*, Springer Theses,
https://doi.org/10.1007/978-3-030-61939-8_1

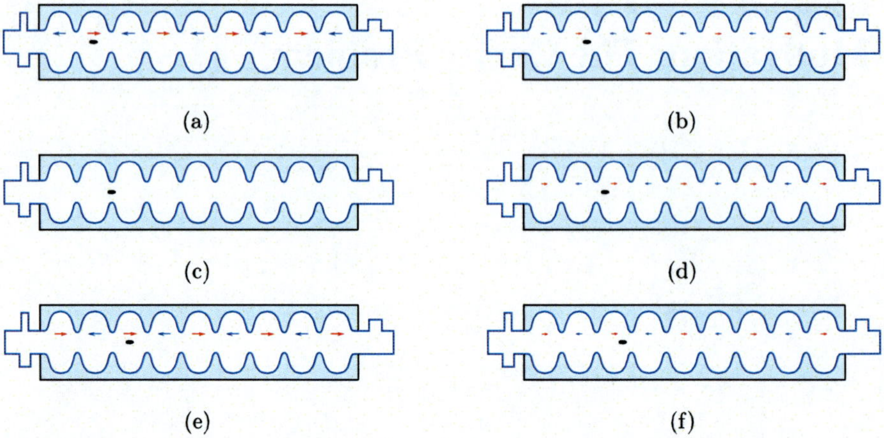

Fig. 1.1 The main principle behind cavities is as follows. The cavity is designed such that within each cell, the electric field accelerates the charged particles forward (**a**). Once the particles leave the cell (**b**), the standing electric wave within the cavity slowly begins to change polarity, until the field is zero while the particle is in between two cells (**c**). Upon entering the next cell (**d**), the direction of the field has changed and the particle is again accelerated forward, where the maximum field strength is reached ideally in the centre of each cell (**e**). This scheme repeats for each cell (**f**), such that in every cell the particle experiences a forward directed force. See also [2]

These requirements are challenging for established numerical methods. Lowest-order schemes such as the *finite integration technique* [4] do not reach the necessary accuracies since their low convergence rates would require the assembly and solution of ill-conditioned systems of sizes which are too large to handle. Moreover, even higher-order schemes struggle since these mostly mesh-based methods introduce errors in the geometry model which are already irreconcilable with the desired error-margin.

This thesis investigates weather it is possible to achieve accuracies of this magnitude via numerical simulations based on isogeometric boundary element methods.

1.1 Isogeometric Boundary Elements for Electromagnetism

This thesis proposes a novel numerical method satisfying the high accuracy requirements and governing assumptions of the cavity problem, namely an *isogeometric boundary element method (IGA BEM)*, based on a discretisation of the electric field integral equation (EFIE). Through the so-called *isogeometric* discretisation [5], parametric mappings are used to describe the geometry. This means that no meshing errors are introduced and that the utilisation of discrete spaces of higher-order is viable.

The proposed isogeometric boundary element method offers doubled rates of convergence with respect to the polynomial degree if the pointwise electric field is the quantity of interest. This convergence behaviour can be exploited to accurately compute the resonant frequencies via so-called *contour integral methods* [6]. These methods are capable of handling the introduced non-linear resonance problem without solving a problem in the vicinity of the resonant frequency, which would introduce instable behaviour.

Since contour integral methods on their own can exhibit exponential convergence, the conducted numerical experiments show the doubled order of convergence of the boundary element method to be preserved in the approximation of the eigenvalue.

These notions are explained in-depth. We show the stability of the isogeometric discretisation away from resonances and, for the first time, prove optimal convergence properties for isogeometric boundary element schemes. All of this is verified through a series of numerical examples, which shows that the high accuracy demands can be satisfied.

1.2 Common Acronyms

See Table 1.1.

1.3 Structure of the Document

We first establish a context for the isogeometric boundary element method, which we discuss in the remainder of this chapter. Afterwards, Chap. 2 discusses the fundamentals on which the approach is built. Chapter 3 introduces and analyses the discretisation scheme, and Chap. 4 addresses a fast implementation thereof. The

Table 1.1 Common Acronyms used throughout this thesis

Acronym	Meaning
ACA	Adaptive cross approximation
BEM	Boundary element method
CAD	Computer aided design
CIM	Contour integral method
DOF	Degree of freedom
FEM	Finite element method
FMM	Fast multipole method
IGA	Isogeometric analysis
IGA-BEM	Isogeometric boundary element method
NURBS	Non uniform rational B-Splines
RT	Raviart–Thomas

numerical verification of the theoretic predictions is conducted in Chap. 5, which reviews a series of numerical examples with the electric field as a quantity of interest. After this, the algorithmic approach which we use to find the resonant frequencies is explained in Chap. 6, which also includes corresponding numerical examples. Appendix A provides a summary of the approximation properties and analogous finite element approximation in three dimensions would exhibit. Lastly, Appendix B provides a brief overview of the numerical implementation that has been written as part of this dissertation project.

1.4 Historical Context

Following Cheng and Cheng [7] and Townsend [8], we will provide some historical context w.r.t. isogeometric boundary element methods. The main steps leading towards the content discussed in this thesis are depicted in Figs. 1.2 and 1.3. Therein, we do not address the history of particle acceleration and Maxwell's equations, since the former is merely a specific motivation for this work and the second is already concisely presented by Rautio [9].

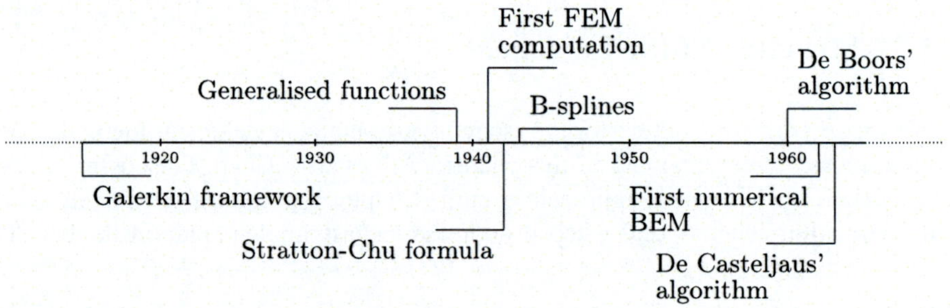

Fig. 1.2 Development before the year 1965

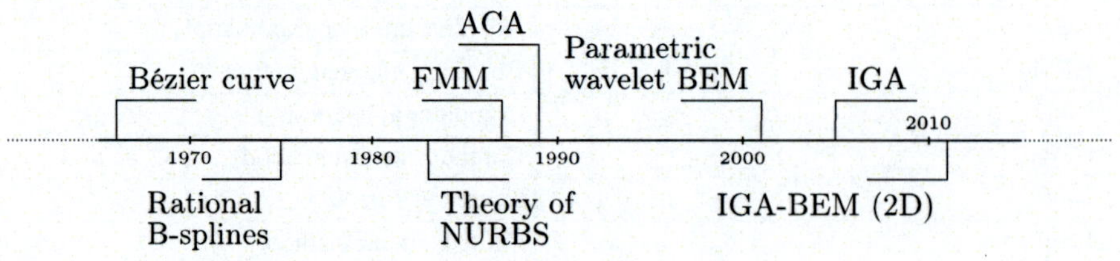

Fig. 1.3 Development after the year 1965

1.4.1 The Foundation of Variational Methods

Since this thesis is exclusively concerned about the Galerkin framework, both the paper of Ritz [10] in 1909 and Galerkin [11] in 1915 should be mentioned. In these, the basis of what is now known as *variational formulation* was introduced. While the foundation for todays numerical framework was laid, a suitable way to handle generalised functions (i.e. Sobolev spaces) was not introduced until Sobolev [12] used introduced notions to investigate the theory of linear hyperbolic differential equations in 1936.

Dissatisfied with purely theoretical assertions, applied mathematicians refined these notions to a point which enabled them to explicitly compute solutions to real world problems. Courant [13] is commmonly attributed with achieving this goal in his 1943 paper, where he uses the variational framework to compute mechanical forces by hand. It was one of the first applications, if not the first, of what has since developed itself into the *finite element method*, see [14]; a method often discretising the domain of interest via small triangles (for two dimensions) or tetrahedrons (for three dimensions) on which computable basis functions are defined. These are then used to represent an approximation of the quantity of interest.

1.4.2 Solution to Boundary Value Problems via Boundary Integrals

While, in its core, the finite element method requires knowledge about the interaction of discrete entities within the domain of interest, scientists were well aware that the data on the boundary, together with some assumptions on the behaviour of the problem, would fix a solution within. The idea that solutions to partial differential equations could be represented by other quantities defined exclusively on the boundary was first formulated under the name *potential theory*. It was investigated by the famous scientists Leonhard Euler (1707–1783), Joseph-Louis Lagrange (1736–1813), Pierre-Simon Laplace (1749–1827) and many more [7].

This eventually led to the first well-known integral identities that relate quantities within the domain with quantities on the boundary; whose development is most commonly attributed to Carl Friedrich Gauß (1777–1855), George Green (1793–1841), and George Gabriel Stokes (1819–1903). The identities derived by Green [15] in 1828 yield exactly the boundary integral equation used to solve electrostatic problems via boundary element methods.

These integral formulae resemble the representation theorems utilised by modern boundary element methods. Derivation of such *representation formulae* continues until today, were the most famous one for the case of electromagnetic problems, the so-called *Stratton–Chu* representation formula, goes back to the early 1940s, cf. [16].

The first numerical solutions of physical problems via boundary integral equations were achieved by multiple groups around the year 1962, cf. [7, Sect. 8]. A collective mentioning of such methods under the name *boundary element method* can be dated back to the year 1977, cf. [7, Sect. 10].

Due to the reformulation of the problem in terms of integral equations, the system matrix becomes densely populated. This made the boundary element method permissively expensive for problems of industrial scope, until the introduction of so called *fast methods*.

Two such fast methods are the *adaptive cross approximation* introduced by Rjasanow [17] in 1998 and the *fast multipole method* introduced by Greengard and Rokhlin [18] in 1987.

1.4.3 Splines and Computational Geometry

Largely disjoint from the increasingly prominent field of numerics of partial differential equations, the mathematician was working at a US military facility [8]. Instead of engineering, he was tasked with the analysis and processing of actuarial data, fitting smooth curves to given data [19].

During this time, he developed the notion of piecewise polynomials of global smoothness (up to a certain degree), which he called splines. The term spline was adapted from smoothly bent wooden planks used in shipbuilding and other constructions, cf. Fig. 1.4. A particular basis was given by the so called *B-splines*, short for basis-splines, for which de Boor [20] derived a well-known recursion formula.

In the 1960s in France, a link to geometry representation was made independently of this development. The mathematician Paul de Faget de Casteljau was hired by Citroën to overcome problems with new manufacturing tools.

De Casteljau was tasked to find a way to handle the free-form curves drawn by Citroën's designers numerically. Indeed, de Casteljau found an algorithmic solution for the task by 1963 [21]. However, this was done in strict secrecy, and de Casteljau was not allowed to publish the findings.

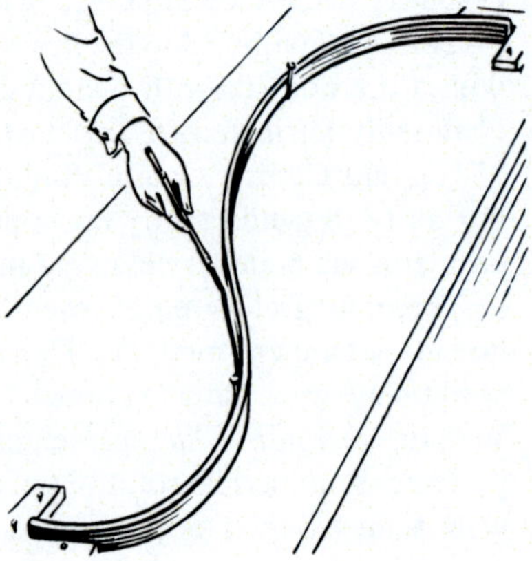

Fig. 1.4 Line-art drawing of a wooden spline as used in shipbuilding. The spline is only fixed at certain points and creates a curvature with minimal bending energy

Although de Casteljaus' work was not made public, competing companies became aware of Citroën's effort. Eventually, word of de Casteljau's achievement reached the head of Renault's design department, a man called Pierre Bézier. Although he knew of the achievement, he did not know of the algorithmic approach. After starting anew, he miraculously came up with a mathematical concept equivalent to de Casteljaus' algorithm. His concept of modeling free-form curves through mathematical equations is today known as *Bézier curves*. This notion is still utilised today in unaltered form to describe two-dimensional vector graphics.

In contrast to Citroën, Renault allowed to publish these results, so the result was accredited to Bézier rather than de Casteljau [8].

De Casteljau's work can be seen as a special case of Schoenbergs initially one-dimensional notion of splines. This notion was later on extended to make the notions even more expressive, first through the construction of rational B-spline curves [22] and then through a generalisation to so called *non-uniform rational B-splines (NURBS)*, cf. Piegl and Tiller [23]. Through these, a large class of geometrical shapes could be described. While Bézier curves are still used to describe vector graphics, NURBS and related notions are so expressive that they are used in modern *computer aided design (CAD)* frameworks to model designs of industrial scope [23, 24].

1.4.4 Linking Design and Simulation

Different communities had already adopted the practice of using geometrically precise parametric mappings for their computations, cf. [25]. However, a thorough link between the geometry and the discrete entities on them was made in 2005 by Hughes et al. [5] through the introduction of *isogeometric analysis (IGA)*.

Building on the results reviewed by Piegl and Tiller [23] and the well established and understood theory of the finite element method, Hughes et al. contrived a method to use not unstructured meshes, but NURBS mappings of modern CAD systems to obtain a description of the domain without the introduction of geometrical errors. Hughes and his group identified this as a way to simplify the industrial workflow, since, through the utilisation of the mappings induced by the CAD systems, preprocessing steps could be skipped, compare Figs. 1.5 and 1.6. These preprocessing steps were previously time consuming, known to take up to 80% of the time of the entire simulation workflow [26].

To discretise the quantity of interest NURBS were utilised as well. Thus, when computing mechanical displacements as quantities of interest, this gave the possibility to immediately deform the geometry. This could be achieved by applying the computed displacements directly to the description of the domain, due to the matching discretisation.

This idea could be seen as an extension of the finite element framework by a more elaborate geometry description and discrete spaces. Nonetheless, it was immensely successful and spawned a whole new area of research, cf. [24, 27–30] and many more. One downside of the isogeometric scheme is that most CAD frameworks operate only in terms of a boundary discretisation, and Hughes' idea required a volumetric

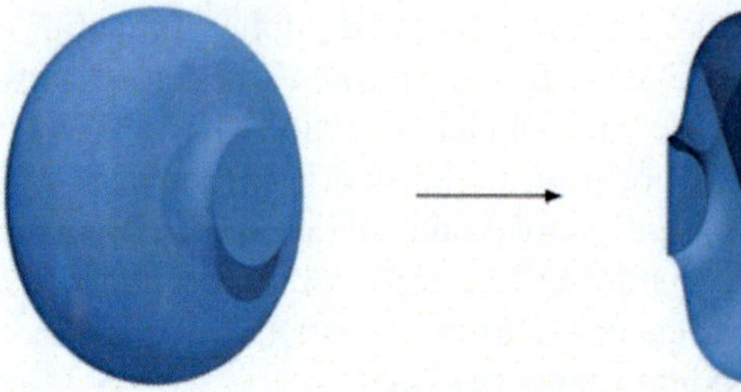

Fig. 1.5 Isogeometric boundary element methods enable the computation directly on the CAD representation

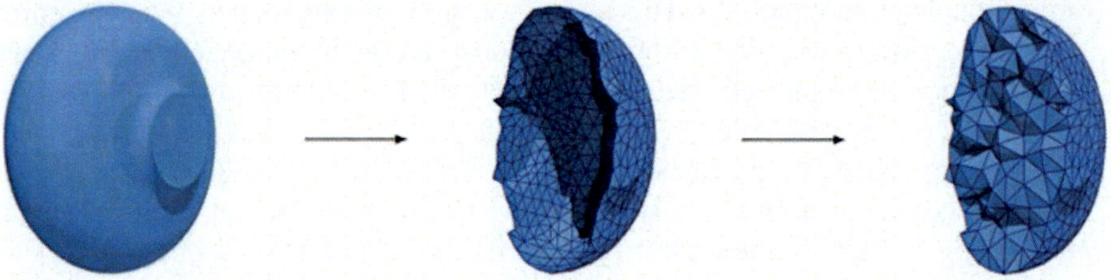

Fig. 1.6 For a FEM computation a mesh is generated from the boundary data available from the design framework. Afterwards, a volume mesh is created

description of the geometry. Thus, a straightforward connection of simulation tools to CAD packages was still challenging. This predominant utilisation of boundary representations in CAD led several engineers to adapt Hughes' ideas to boundary element methods, which avoided the need for the creation of a volume discretisation. The first known numerical results were published in 2012 by Simpson et al. [31] and Takahashi and Matsumoto [32], dealing with simple two-dimensional examples.

1.5 State-of-the-Art and Contribution

We will now present a review of the state-of-the-art, together with an explanation of the authors' contribution.

1.5.1 Volume-Based Methods

As of today, volume-based methods for the simulation of electromagnetic problems can be regarded as the industry standard. Most implementations rely on conforming discretisations, which in the context of finite element methods exist both for classical as well as for isogeometric formulations and are well understood, see [29, 33,

34]. Another volume-based method, which is highly established within the field of computational electromagnetism, is the *finite integration technique*, see the works of Weiland et al. [4] and the references cited therein. As mentioned before, all of these volume-based methods require pre-processing to obtain a volumetric discretisation, which will inevitably introduce some errors. Moreover, automated generation of parametrisations for volumes is an open area of research.

1.5.2 Non-volumetric Methods

For the solution of electromagnetic problems with a boundary representation alone, classical boundary element methods are a valid choice. Three sound implementations are discussed by Hiptmair and Kielhorn [36], Śmigaj et al. [37], and Weggler [35]. Of these only the code of Weggler [35] is capable of higher-order simulations, and only the implementation of Śmigaj et al. [37] is freely available. Moreover, each of these approaches is mesh-based and, once again, suffers from the need of a geometry discretisation, introducing geometrical errors. An approach via wavelets is also possible, cf. [25], and will be discussed below.

A non-isogeometric method minimising the impact of geometry discretisations was introduced by Li et al. [38]. The paper presents an approach to the electric field integral equation which is based on *subdivision*, a scheme going back to the work of Catmull and Clark [39]. There, the exact geometry and the discrete quantities are given as the limit of an exponentially converging recursive formula, which yields a refined mesh in each iteration. Subdivision algorithms can be implemented more efficiently than algorithms based on B-spline evaluations. Moreover, subdivision schemes offer the possibility to use globally smooth discrete spaces, which generally cannot be achieved by isogeometric methods due to patch boundaries in multipatch domains. However, they do not pose a flawless alternative to the isogeometric approach. First, in real-world examples, globally smooth geometries do not play important roles, strongly weakening the main advantage advertised. Secondly, most CAD frameworks operate with (trimmed) NURBS as boundary representation, while subdivision surfaces are mostly used for animations. Thus, before subdivision schemes are applicable, preprocessing steps must be taken. This does not make subdivision-based approaches attractive for the inclusion into manufacturing workflows since they do not offer the user-friendliness of the isogeometric approach.

1.5.3 Isogeometric Boundary Element Methods

Simpson et al. [31] had already implemented an IGA-BEM for two-dimensional problems, and a simple implementation of a collocation-based approach for three-dimensional problems was discussed in [40]. However, the first isogeometric three-dimensional Galerkin BEM was presented in our paper [41]. There, we introduce a fast isogeometric discretisation of single layer approaches to both Laplace equation and Helmholtz equation. Moreover, for the first time, we discussed the provably optimal behaviour of the interpolation-based multipole compression for scalar problems as well as analytic stability of the isogeometric discretisation.

With regards to electromagnetism, Simpson et al. [42] published numerical examples of a three-dimensional IGA-BEM for the electric field integral equation. However, they do not address the electric field as a quantity of interest and do not make any analytical considerations related to compression and the stability of the discretisation. These questions were first discussed by us in [43], where we present the first provably optimal compression scheme for isogeometric discretisations of boundary element methods for electromagnetism, a proof of inf-sup stability of the isogeometric EFIE, and a series of numerical examples, including a detailed investigation of the pointwise error in the electric field.

A comparison of the isogeometric ansatz spaces to different discretisations of the electric field integral equation was first conducted by us in [44, 45], where we compare the spline-based discretisation to parametric Raviart–Thomas elements and also discuss spectral elements, respectively. Our implementation used for the numerical studies of [43–45] was made publicly available [46], archived [47], and its API is explained in [48]. To the best of the authors' knowledge, this is the only publicly available isogeometric boundary element code, and the only publicly available code for the EFIE capable of higher-order computations.

1.5.4 Fast Methods

Within the spline-based isogeometric boundary element framework, one could also consider a different choice of a fast method. Different multipole methods are available, see for example [49, 50], but are rarely black-box. The method used in this thesis is based on the previous work of Dölz et al. [51] and has also been applied to other PDEs [41]. An alternative to multipole methods is given by the so-called adaptive cross approximation (ACA) [52]. However, a comparison to other fast methods conducted by Harbrecht and Peters [53] favoured multipole-based compression techniques over the ACA on parametric surfaces for lowest-order discretisations. An account of a promising ACA specific to the isogeometric setting can be found in [54], however, not in the context of electromagnetism.

One could also utilise a boundary element approach via wavelets, cf. [25, 55], which was the most favourable approach in the above-mentioned comparison [53]. While the results remain largely unknown to engineering communities, the theoretical properties of this approach are understood extremely well, see the works of Dahmen et al. [56]. The wavelet approach for boundary element methods also does not suffer from the need for fast methods. Due to a property of wavelet bases called *vanishing moments* the system matrices obtained from this approach are almost sparse, in the sense that a large number of entries are small enough to be considered zero. Thus, a fast boundary element approach via wavelets yields sparse systems. However, implementation of wavelets is more involved than the implementation of the isogeometric BEM proposed in this thesis, and, to the best of the authors' knowledge, no studies of higher-order divergence-conforming wavelets have been conducted.

1.5.5 Approximation Properties

The analysis of the isogeometric boundary element approach is highly dependent on suitable approximation estimates for the spline spaces. For domains diffeomorphic to the unit cube, so-called *single patch domains*, these were introduced by Buffa et al. [34]. A generalisation to other domains, so-called *multipatch domains* was presented for globally continuous scalar functions in [57]. The non-trivial generalisation of this concept to curl and div-conforming spaces, as well as quasi-optimal error estimates w.r.t. the spaces $H^{-1/2}(\Gamma)$, $\boldsymbol{H}_{\times}^{-1/2}(\mathrm{div}_\Gamma, \Gamma)$, and $H^{1/2}(\Gamma)$ for multipatch boundaries Γ were introduced by us in [58].

1.5.6 Electromagnetic Resonance Problems in the Context of IGA and BEM

As made clear in the motivation, resonance problems are of specific interest. An approach to the linear eigenvalue problem via isogeometric finite elements was presented in [29].

However, an approach via boundary element methods renders the electromagnetic eigenvalue problem non-linear. A study for the corresponding boundary element approach has only been considered outside the isogeometric setting [59], where a Newton method is used to solve the non-linear eigenvalue problem. Indeed, an algorithmic approach to non-linear eigenvalue problems was difficult, see e.g. the overview of Mehrmann and Voss [60]. This changed with the introduction of a new class of algorithms in 2012, the so-called *contour integral methods*, independently presented by Asakura et al. [61] and Beyn [6]. A comparison of multiple approaches was conducted by Imakura et al. [62]. There, the authors concluded that all of the considered contour integral methods can be regarded as projection methods and can be categorised based on the subspaces projected on, as well as on the type of projections and the problem to which they are applied implicitly.

The first approach via a contour integral method together with an isogeometric BEM is presented in this thesis. For classical lowest-order BEM and without a fast method this has already been studied numerically by Elasmi [63]. The analysis of the contour integral method in conjunction with general Galerkin boundary element methods based on the electric field integral equation has recently been presented by Unger [64].

References

1. Aune B, Bandelmann R, Bloess D, Bonin B, Bosotti A, Champion M, Crawford C, Deppe G, Dwersteg B, Edwards DA, Edwards HT, Ferrario M, Fouaidy M, Gall P-D, Gamp A, Gössel A, Graber J, Hubert D, Hüning M, Juillard M, Junquera T, Kaiser H, Kreps G, Kuchnir M, Lange R, Leenen M, Liepe M, Lilje L, Matheisen A, Möller W-D, Mosnier A, Padamsee H, Pagani C, Pekeler M, Peters H-B, Peters O, Proch D, Rehlich K, Reschke D, Safa H, Schilcher T, Schmuser P, Sekutowicz J, Simrock S, Singer W, Tigner M, Trines D, Twarowski K, Weichert G, Weisend J, Wojtkiewicz J, Wolff S, Zapfe K (2000) Superconducting TESLA cavities. Phys Rev Accel Beams 3(9):092001
2. Wiedemann H (2007) Particle accelerator physics, 3rd edn. Springer, Berlin
3. Georg N, Ackermann W, Corno J, Schöps S (2019) Uncertainty quantification for Maxwell's eigenproblem based on isogeometric analysis and mode tracking. Comput Methods Appl Mech Eng 350:228–244
4. Weiland T, Timm M, Munteanu I (2008) A practical guide to 3-D simulation. IEEE Microwave Mag 9(6):62–75
5. Hughes TJR, Cottrell JA, Bazilevs Y (2005) Isogeometric analysis: CAD, finite elements, NURBS, exact geometry and mesh refinement. Comput Methods Appl Mech Eng 194:4135–4195
6. Beyn W-J (2012) An integral method for solving nonlinear eigenvalue problems. Linear Algebra Appl 436(10):3839–3863. Special Issue dedicated to Heinrich Voss's 65th birthday
7. Cheng AH-D, Cheng DT (2005) Heritage and early history of the boundary element method. Eng Anal with Bound Elem 29(3):268–302
8. Townsend A (2014) On the spline: a brief history of the computational curve. Int J Inter Arch + Spatial Design 2(3):48–59
9. Rautio J (2014) The long road to Maxwell's equations. IEEE Spect 51(12):36–56
10. Ritz W (1909) Über eine neue Methode zur Lösung gewisser Variationsprobleme der mathematischen Physik. Journal für reine und angewandte Mathematik 135:1–61
11. Galerkin BG (1915) Rods and plates. Series occurring in various questions concerning the elasticequilibrium of rods and plates. Engineers Bulletin (Vestnik Inzhenerov) 19:897–908 In Russian
12. Sobolev SL (1936) Méthode nouvelle à résoudre le problème de Cauchy pour les équations linéaires hyperboliques normales. Matematicheskiĭ Sbornik 1 43(1):39–72
13. Courant R (1943) Variational methods for the solution of problems of equilibrium and vibrations. Bull Amer Math Soc 49:1–23
14. Braess D (2007) Finite elemente. Springer, Berlin
15. Green G (1828) An essay on the application of mathematical analysis to the theories of electricity and magnetism. George Green, Göteborg
16. Stratton JA (1941) Electromagnetic theory. IEEE Press, Piscataway
17. Rjasanow S (1998) The structure of the boundary element matrix for the threedimensional Dirichlet problem in elasticity. Numer Linear Algebra Appl 5(3):203–217
18. Greengard L, Rokhlin V (1987) A fast algorithm for particle simulations. J Comput Phys 73(2):325–348
19. Schoenberg IJ (1946) Contributions to the problem of approximation of equidistant data by analytic functions, Parts A and B. Appl Math 4:45–99
20. de Boor C (2001) A practical guide to splines, vol 27, Revised edn. Applied mathematical sciences. Springer, New York
21. de Casteljau PdF (1963) Courbes et surfaces à Pôles. Technical Report, Paris, Citröen
22. Piegl L, Tiller W (1997) The NURBS book, 2nd edn. Springer, Berlin
23. Piegl L, Tiller W (1987) Curve and surface constructions using rational B-splines. Comput-Aided Design 19(9):485–498
24. Cottrell JA, Hughes TJR, Bazilevs Y (2009) Isogeometric analysis: toward integration of CAD and FEA. Wiley, New York

25. Harbrecht H (2011) Wavelet Galerkin schemes for the boundary element method in three dimensions. Dissertation. Technische Universität Chemnitz
26. Boggs PT, Althsuler A, Larzelere AR, Walsh EJ, Clay RL, Hardwick MF (2005) DART system analysis. Technical report, Sandia National Laboratories
27. Bontinck Z, Corno J, De Gersem H, Kurz S, Pels A, Schöps S, Wolf F, de Falco C, Dölz J, Vázquez R, Römer U (2017) Recent advances of isogeometric analysis in computational electromagnetics. In: ICS Newsletter (International Compumag Society) 3. arXiv:1709.06004
28. Buffa A, Sangalli G, Vázquez R (2013) Isogeometric methods for computational electromagnetics: B-spline and T-spline discretizations. J Comput Phys 257(Part B.0):1291–1320
29. Corno J, de Falco C, De Gersem H, Schöps S (2016) Isogeometric simulation of Lorentz detuning in superconducting accelerator cavities. Comput Phys Commun 201:1–7
30. Beer G, Mallardo V, Ruocco E, Marussig B, Zechner J, Dunser C, Fries T-P (2017) Isogeometric boundary element analysis with elasto-plastic inclusions. Part 2: 3-D problems. Comput Methods Appl Mech Eng 315(Supplement C):418–433
31. Simpson RN, Bordas S, Trevelyan J, Rabczuk T (2012) A two-dimensional isogeometric boundary element method for elastostatic analysis. Comput Methods Appl Mech Eng 209(212):87–100
32. Takahashi T, Matsumoto T (2012) An application of fast multipole method to isogeometric boundary element method for Laplace equation in two dimensions. Eng Anal Bound Elem 36(12):1766–1775
33. Monk P (2003) Finite element methods for Maxwell's equations. Oxford University Press, Oxford
34. Buffa A, Sangalli G, Vázquez R (2010) Isogeometric analysis in electromagnetics: B-splines approximation. Comp Methods Appl Mech Eng 199:1143–1152
35. Weggler L (2011) High order boundary element methods. Dissertation. Universitat des Saarlandes, Saarbrücken
36. Hiptmair R, Kielhorn L (2012) BETL-a generic boundary element template library. Technical Report, Seminar for Applied Mathematics, ETH Zurich, pp 2012–36
37. Śmigaj W, Betcke T, Arridge S, Phillips J, Schweiger M (2015) Solving boundary integral problems with BEM++. ACM Trans Math Softw 41(2):1–40
38. Li J, Dault D, Zhao R, Liu B, Tong Y, Shanker B (2015) Isogeometric analysis of integral equations using subdivision. In: 2015 IEEE international symposium on antennas and propagation USNC/URSI national radio science meeting, pp 153–154
39. Catmull E, Clark J (1978) Recursively generated B-spline surfaces on arbitrary topological meshes. Comput Aided Design 10(6):350–355
40. Simpson R, Scott M, Taus M, Thomas D, Lian H (2014) Acoustic isogeometric boundary element analysis. Comput Methods Appl Mech Eng 269:265–290
41. Dölz J, Harbrecht H, Kurz S, Schöps S, Wolf F (2018) A fast isogeometric BEM for the three dimensional Laplace- and Helmholtz problems. Comput Methods Appl Mech Eng 330(Supplement C):83–101
42. Simpson R, Liu Z, Vázquez R, Evans J (2018) An isogeometric boundary element method for electromagnetic scattering with compatible B-spline discretizations. J Comput Phys 362:264–289
43. Dölz J, Kurz S, Schöps S, Wolf F (2019) Isogeometric boundary elements in electromagnetism: rigorous analysis, fast methods, and examples. SIAM J Sci Comput 41(5):B983–B1010
44. Kurz S, Schöps S, Wolf F (2019) Towards a spectral method of moments using computer aided design
45. Dölz J, Kurz S, Schöps S, Wolf F (2019) A numerical comparison of an isogeometric and a parametric higher order Raviart-Thomas approach to the electric field integral equation. IEEE Trans Antennas Propag 68(1):593–597
46. Dölz J, Harbrecht H, Kurz S, Multerer M, Schöps S, Wolf F (2019) Bembel. The C++ boundary element engineering library. Official website, www.bembel.eu. Accessed 20 Aug 2019
47. Dölz J, Harbrecht H, Kurz S, Multerer M, Schops S, Wolf F (2019) Bembel. The C++ boundary element engineering library. Archived Version 0.9. Zenodo. https://doi.org/10.5281/zenodo.2671596

48. Dölz J, Harbrecht H, Kurz S, Multerer M, Schöps S, Wolf F (2020) Bembel: the fast isogeometric boundary element C++ library for laplace, Helmholtz, and Electric Wave Equation. In: SoftwareX 11, p 10476
49. Schobert DT, Eibert TF (2010) A multilevel interpolating fast integral solver with fast Fourier transform acceleration. In: 2010 URSI international symposium on electromagnetic theory, pp 520–523
50. Schobert DT, Eibert TF (2012) Fast integral equation solution by multilevel Green's function interpolation combined with multilevel fast multipole method. IEEE Trans Antennas Propag 60(9):4458–4463
51. Dölz J, Harbrecht H, Peters M (2016) An interpolation-based fast multipole method for higher-order boundary elements on parametric surfaces. Int J Numer Methods Eng 108(13):1705–1728
52. Hackbusch W (2015) Hierarchical matrices: algorithms and analysis. Springer, Berlin
53. Harbrecht H, Peters M (2013) Comparison of fast boundary element methods on parametric surfaces. Comput Methods Appl Mech Eng 261:39–55
54. Marussig B, Zechner J, Beer G, Fries T-P (2015) Fast isogeometric boundary element method based on independent field approximation. Comput Methods Appl Mech Eng 284:458–488. Isogeometric Analysis Special Issue
55. Jensen TK, Christensen O, Pedersen M (2003) On adaptive wavelet-based methods for the Maxwell equations. Technical University of Denmark, Kongens Lyngby, Denmark
56. Dahmen W, Harbrecht H, Schneider R (2006) Compression techniques for boundary integral equations. Asymptotically optimal complexity estimates. SIAM J Numer Anal 43(6):2251–2271
57. Buffa A, Vázquez RH, Sangalli G, da Veiga LB (2015) Approximation estimates for isogeometric spaces in multipatch geometries. Numer Methods Partial Diff Equ 31(2):422–438
58. Buffa A, Dölz J, Kurz S, Schöps S, Vázquez R, Wolf F (2020) Multipatch approximation of the de Rham sequence and its traces in isogeometric analysis. Numerische Mathematik 144:201–236
59. Xin J (2011) Boundary element approximation for Maxwell's eigenvalue problem. Dissertation. Karlsruhe Institute of Technology, Karlsruhe, Germany
60. Mehrmann V, Voss H (2004) Nonlinear eigenvalue problems: a challenge for modern eigenvalue methods. GAMM-Mitteilungen 27(2):121–152
61. Asakura J, Sakurai T, Tadano H, Ikegami T, Kimura K (2009) A numerical method for nonlinear eigenvalue problems using contour integrals. JSIAM Lett 1:52–55
62. Imakura A, Du L, Sakurai T (2016) Relationships among contour integral-based methods for solving generalized eigenvalue problems. Jpn J Ind Appl Math 33(3):721–750
63. Elasmi M (2017) Boundary element method for Maxwell's eigenvalue problems in TESLA cavities using the contour integral method. Master's Thesis. Universität des Saarlandes, Saarbrücken
64. Unger G (2017) Convergence analysis of a Galerkin boundary element method for electromagnetic eigenvalue problems. Technical Report 2017/2, Institute of Computational Mathematics, Graz University of Technology

Chapter 2
Foundations

From now on, in order to avoid the repeated use of generic but unspecified constants, by $C \lesssim D$ we mean that C can be bounded by a multiple of D, independently of parameters fixed through context or which D may depend on. In the usual sense, $C \gtrsim D$ is defined as $D \lesssim C$, and $C \simeq D$ as $C \lesssim D \lesssim C$.

2.1 Hilbert Spaces

Before we precisely characterise the problems we want to solve, we need to recall some essential functional analytic notions. We follow the books by McLean [1] and Wloka [2], deviating only to cover assertions required for the vector valued cases specific to electromagnetic problems. For this, our main reference is a review article by Buffa and Hiptmair [3].

Definition 2.1 (*Hilbert Space,* [1, p. 38]) Any vector space X equipped with an inner product $\langle\cdot,\cdot\rangle_X$ is said to be a *Hilbert space*, if it is complete w.r.t. the induced norm given by

$$\|x\|_X = \sqrt{\langle x, x\rangle_X} \tag{2.1}$$

for all $x \in X$.

Now we cite a first fundamental result.

Lemma 2.2 (Cauchy–Schwarz Inequality, [1, Eq. (2.5)]) *Let X be a Hilbert space with scalar product $\langle\cdot,\cdot\rangle_X$ and induced norm $\|\cdot\|_X$. Then, for any $x, y \in X$ it holds that*

© The Author(s), under exclusive license to Springer Nature Switzerland AG 2021

F. Wolf, *Analysis and Implementation of Isogeometric Boundary Elements for Electromagnetism*, Springer Theses,
https://doi.org/10.1007/978-3-030-61939-8_2

$$|\langle x, y\rangle_X| \leq \|x\|_X \|y\|_X.$$

This general setting alone already enables us to review some well-known results.

Definition 2.3 (*Gelfand Triple,* [2, Theorem 17.3]) Let X and Y be Hilbert spaces with a dense, continuous, and injective embedding $\mathsf{e}\colon X \hookrightarrow Y$. Defining the *dual norm*

$$\|y\|_{X'} := \sup_{0 \neq x \in X} \frac{|\langle y, \mathsf{e}x\rangle_Y|}{\|x\|_X}, \qquad y \in Y, \tag{2.2}$$

we denote the completion of Y w.r.t. $\|\cdot\|_{X'}$ as X'. The tuple (X, Y, X') is called *Gelfand Triple*, and we say that X' is the *dual space of* X w.r.t. the so-called *pivot space* Y.

At times, to highlight this relationship, we will write $\langle \cdot, \cdot\rangle_{X' \times X}$ instead of $\langle \cdot, \cdot\rangle_Y$.

Remark 2.4 Gelfand triples (X, Y, X') are usually defined for X not being a Hilbert space, but rather in the more general setting of (reflexive) Banach spaces. This generalisation is of no interest for this thesis, thus we choose to simplify the setting.

Proposition 2.5 (Canonical Embedding of the Dual, [2, Theorem 17.3]) *The embedding* $\mathsf{e}'\colon Y \hookrightarrow X'$ *induced by the completion in Definition 2.3 is dense, continuous, and injective. Moreover,* X' *is a Hilbert space.*

Summarising, this yields the density, continuity, and injectivity of the embeddings

$$X \overset{\mathsf{e}}{\hookrightarrow} Y \overset{\mathsf{e}'}{\hookrightarrow} X'.$$

The embeddings utilised within this thesis are exclusively those induced by completion w.r.t. the norms. Thus, we will refrain from using the notation e and e' from this point forward.

By continuity arguments, one can show the following.

Lemma 2.6 (Continuity of the Duality Product, [2, Theorem 17.2]) *Let* (X, Y, X') *be a Gelfand triple. For all* $x \in X$ *and* $y \in X'$ *it holds that*

$$|\langle x, y\rangle_Y| \leq \|x\|_X \|y\|_{X'}.$$

2.2 Approximations in Hilbert Spaces

Having defined the essential notions of Hilbert spaces, we can now review some basic approximation results.

Theorem 2.7 (Best Approximation in Hilbert Spaces, [1, Lemma 2.28]) *Let X be a Hilbert space and let $Y \subseteq X$ be a nonempty, closed, and convex subset. Then for any $u \in X$ there exists a unique element $\tilde{u} \in Y$ such that*

$$\|\tilde{u} - u\|_X = \inf_{y \in Y} \|y - u\|_X.$$

Theorem 2.7 particularly is applicable to any finite dimensional subspace $Y := X_h$ of a Hilbert space X, since it fulfils the closedness and convexity assumption by their definition via the span of a finite basis. These best approximations can be obtained by orthogonal projections.

Definition 2.8 (*Orthogonal Projection,* [1, p. 40]) Let X be a Hilbert space and let $Y \subseteq X$ be a nonempty, closed, and convex subset. We define the *X-orthogonal projection* $\mathcal{P}: X \to Y$ such that $\mathcal{P}u = \tilde{u}$, where $\tilde{u} \in Y$ is the solution to the problem of Theorem 2.7.

Indeed, for any orthogonal projection $\mathcal{P}$ it is clear that $\mathcal{P}$ is a projection in the sense of $\mathcal{P} \circ \mathcal{P} = \mathcal{P}$. This projection induces the *orthogonal decomposition*

$$X = Y \oplus Y^{\intercal},$$

where $Y = \operatorname{im} \mathcal{P}$ and $Y^{\intercal} = \ker \mathcal{P}$. Commonly, $Y^{\intercal}$ is referred to as *orthogonal complement of Y in X*.

We now introduce three essential results discussing well-posedness of variational problems and the quality of approximate solutions.

The first result is due to Céa [4], and may also be found in many standard references to the finite element method, e.g. the book by Monk [5].

Theorem 2.9 (Céa's Lemma, [5, p. 25]) *Let X be a Hilbert space endowed with the norm $\|\cdot\|_X$. Let $a: X \times X \to \mathbb{K}$ be a continuous and X-coercive sesquilinear form over X and let $\ell: X \to \mathbb{K}$ be a bounded linear operator. Let $u \in X$ denote the solution to the problem*

$$a(u, v) = \ell(v), \qquad \textit{for all } v \in X.$$

Then, for any finite dimensional space $X_h \subseteq X$ the solution $u_h \in X_h$ to the problem

$$a(u_h, v_h) = \ell(v_h), \qquad \textit{for all } v_h \in X_h,$$

satisfies

$$\|u - u_h\|_X \lesssim \min_{v_h \in X_h} \|u - v_h\|.$$

Theorem 2.10 (Inf-Sup Condition, [6]) *Let U, V be Hilbert spaces, and let $a: U \times V \to \mathbb{K}$ be a continuous bilinear form with $a(u, v) \lesssim \|u\|_U \|v\|_V$, for all $u \in U$, $v \in V$. Then the variational problem of finding a function $u \in U$ with*

$$a(u, v) = \ell(v), \qquad \textit{for all } v \in V,$$

and an $\ell \in V'$ *is well posed if and only if there exists an* $\alpha \in \mathbb{R}$ *such that*

$$\inf_{0 \neq u \in U} \sup_{0 \neq v \in V} \frac{a(u, v)}{\|u\|_U \|v\|_V} = \inf_{0 \neq v \in V} \sup_{0 \neq u \in U} \frac{a(u, v)}{\|u\|_U \|v\|_V} = \alpha > 0$$

holds.

Theorem 2.11 (Discrete Inf-Sup Condition, [7, Theorem 1]) *Let the assumptions of Theorem 2.10 be satisfied, and let* $U_h \subset U$ *and* $V_h \subset V$ *be two non-trivial subspaces. If the condition*

$$\inf_{0 \neq u_h \in U_h} \sup_{0 \neq v_h \in V_h} \frac{a(u_h, v_h)}{\|u_h\|_U \|v_h\|_V} = \inf_{0 \neq v_h \in V} \sup_{0 \neq u_h \in U} \frac{a(u_h, v_h)}{\|u_h\|_U \|v_h\|_V} = \alpha_h > \epsilon > 0$$

for some $\epsilon > 0$ *is fulfilled, then the variational problem of finding an* $u_h \in U_h$ *with*

$$a(u_h, v_h) = \ell(v_h), \qquad \textit{for all } v_h \in V_h,$$

and an $\ell \in V'$ *is well posed. Furthermore, for* u *being the solution to the problem in Theorem 2.10, it holds that*

$$\|u - u_h\|_U \lesssim \inf_{w_h \in U_h} \|u - w_h\|_U.$$

2.3 Sobolev Spaces in Domains

We are interested in a special class of Hilbert spaces, namely those Hilbert spaces which come in the form of *Sobolev spaces*, i.e., Hilbert spaces of functions. While general L^p-theory is relevant in many areas, we only recall the required theory for this thesis. Let $\Omega \subseteq \mathbb{R}^d$, $d = 2, 3$. For the remainder of the thesis, for any set $\Omega \in \mathbb{R}^d$ we will set $\Omega^c := \mathbb{R}^d \setminus \Omega$.

Definition 2.12 (*C^k-Domain*, [1, p. 89ff]) Let $k \geq 0$. A domain $\Omega \subset \mathbb{R}^d$ with compact boundary Γ is called *C^k -Domain* if

- for a suitable index set J there exists a finite open covering $\{W_j\}_{j \in J}$ of Γ and $\{\Omega_j\}_{j \in J}$ of Ω such that each Ω_j can be transformed to a *C^k-hypograph*, i.e., set of the form

$$\Omega_j = \{\boldsymbol{x} = (x_1, \ldots, x_d) \in \mathbb{R}^d : x_d < \theta_j(\boldsymbol{x}') \text{ for all } \boldsymbol{x}' = (x_1, \ldots, x_{d-1}) \in \mathbb{R}^{d-1}\}$$

  for some C^k functions $\{\theta_j\}_{j \in J}$, by a rigid motion, i.e., rotation and translation, and
- one finds $W_j \cap \Omega_j = W_j \cap \Omega$ for all $j \in J$.

If the θ_j are (only) Lipschitz continuous, we refer to Ω as a *Lipschitz domain* and to the Ω_j as *Lipschitz hypographs*.

Without loss of generality we will omit the inclusion of the rigid motion mentioned above in the following derivations. Moreover, we will call a d-dimensional manifold Γ *Lipschitz boundary* if there exists a Lipschitz domain $\Omega \in \mathbb{R}^{d+1}$ with $\partial\Omega = \Gamma$.

As commonly done, we define the scalar product

$$\langle u, v\rangle_{L^2(\Omega)} = \int_\Omega u(x)\overline{v(x)}\,\mathrm{d}x$$

for all functions $u, v\colon \Omega \to \mathbb{C}$ in which the right hand side is well defined in the sense of the Lebesgue integral. As in Definition 2.1, we can now define the Hilbert space of functions $L^2(\Omega)$ as the closure of the infinitely differentiable functions $C^\infty(\Omega)$ w.r.t. the induced norm

$$\|u\|_{L^2(\Omega)} := \sqrt{\langle u, v\rangle_{L^2(\Omega)}}. \tag{2.3}$$

To define Sobolev spaces of higher regularity, we first need to introduce multi-index notation. A multi-index is a tuple $\boldsymbol{\alpha}=(\alpha_1, \dots, \alpha_d) \in \mathbb{N}^d$ and $|\boldsymbol{\alpha}| = \sum_{1\le j\le d} \alpha_j$. Moreover, we define the notation $D^\alpha f := \partial_{x_1}^{\alpha_1} \cdots \partial_{x_d}^{\alpha_d} f$ for any sufficiently differentiable function f.

Definition 2.13 (*Sobolev Spaces,* [1, p. 73ff]) Let Ω be a Lipschitz domain. Fixing the notation $H^0(\Omega) = L^2(\Omega)$, we define for any integer $m > 0$ the space of functions $H^m(\Omega)$ as the subset of $L^2(\Omega)$ for which the norm

$$\|u\|^2_{H^m(\Omega)} = \|u\|^2_{H^{m-1}(\Omega)} + \sum_{|\boldsymbol{\alpha}|=m} \left\|D^{\boldsymbol{\alpha}} u\right\|^2_{L^2(\Omega)}$$

induced by the scalar product

$$\langle u, v\rangle_{H^m(\Omega)} = \langle u, v\rangle_{H^{m-1}(\Omega)} + \sum_{|\boldsymbol{\alpha}|=m} \langle D^{\boldsymbol{\alpha}} u, D^{\boldsymbol{\alpha}} v\rangle_{L^2(\Omega)}$$

is finite. For $\Omega \subset \mathbb{R}^d$, $\epsilon \in (0, 1)$ and integers $m \geq 0$ one can define the scalar product $\langle u, v\rangle_{H^{m+\epsilon}(\Omega)}$ as

$$\langle u, v\rangle_{H^m(\Omega)} + \sum_{|\alpha|=m} \int_\Omega\int_\Omega \frac{(\partial^\alpha u(x) - \partial^\alpha u(y))\overline{(\partial^\alpha v(x) - \partial^\alpha v(y))}}{|x-y|^{d+2\epsilon}}\,\mathrm{d}x\,\mathrm{d}y,$$

thus inducing the Sobolev spaces $H^{m+\epsilon}(\Omega)$.

For unbounded domains we will denote any Sobolev space $H^s_{\mathrm{loc}}(\Omega^c)$ for $s \geq 0$ as the space of functions f for which $f \in H^s(D)$ holds for any precompact domain $D \subsetneq \Omega^c$.

Of specific importance are spaces of type $H^1(\Omega)$. Thus, for ease of notation, we denote their seminorm term by $|\cdot|_{H^1(\Omega)}$. It can be related to the $L^2(\Omega)$-norm as follows.

Lemma 2.14 (Poincaré Inequality, [2, Theorem 7.7]) *Let $\Omega \subset \mathbb{R}^d$ be a compact and connected Lipschitz domain with Lebesgue measure $|\Omega|$. Let, moreover, $u \in H^1(\Omega)$. Then, for $\tilde{u} := |\Omega|^{-1} \int_\Omega u(\boldsymbol{x})\,\mathrm{d}\boldsymbol{x}$, there exists a constant $C = C(\Omega)$ such that*

$$\|u - \tilde{u}\|_{L^2(\Omega)} \leq C|u|_{H^1(\Omega)}.$$

All of these notions can be extended to vector-valued Sobolev spaces, where the regularity assumptions need to be understood component-wise. Thus, we set

$$\boldsymbol{L}^2(\Omega) := \left(L^2(\Omega)\right)^d \quad \text{and} \quad \boldsymbol{H}^s(\Omega) := \left(H^s(\Omega)\right)^d.$$

Note that these spaces still maintain a Hilbert space structure, since the respective norms can be induced by defining corresponding scalar products via summation of the products for scalar-valued functions of each vector component.

On Lipschitz domains one last class of function spaces need to be introduced. For $s \geq 0$ we define the scalar products

$$\langle u, v\rangle_{\boldsymbol{H}^s(\mathbf{curl},\Omega)} := \langle u, v\rangle_{\boldsymbol{H}^s(\Omega)} + \langle \mathbf{curl}\, u, \mathbf{curl}\, v\rangle_{\boldsymbol{H}^s(\Omega)}, \quad \text{and}$$
$$\langle u, v\rangle_{\boldsymbol{H}^s(\mathrm{div},\Omega)} := \langle u, v\rangle_{\boldsymbol{H}^s(\Omega)} + \langle \mathrm{div}\, u, \mathrm{div}\, v\rangle_{H^s(\Omega)}.$$

These induce the norms

$$\|\boldsymbol{u}\|_{\boldsymbol{H}^s(\mathbf{curl},\Omega)} := \sqrt{\|\boldsymbol{u}\|^2_{\boldsymbol{H}^s(\Omega)} + \|\mathbf{curl}\boldsymbol{u}\|^2_{\boldsymbol{H}^s(\Omega)}}, \quad \text{and}$$
$$\|\boldsymbol{u}\|_{\boldsymbol{H}^s(\mathrm{div},\Omega)} := \sqrt{\|\boldsymbol{u}\|^2_{\boldsymbol{H}^s(\Omega)} + \|\mathrm{div}\,\boldsymbol{u}\|^2_{H^s(\Omega)}}.$$

We set the corresponding spaces $\boldsymbol{H}^s(\mathbf{curl}, \Omega)$ and $\boldsymbol{H}^s(\mathrm{div}, \Omega)$ as the subspaces of $\boldsymbol{L}^2(\Omega)$ for which the respective norm is finite. These spaces will be referred to as spaces with *graph norm*. Note that this notion can be extended to further differential operators in complete analogy, specifically to $\mathbf{curl}^2 = \mathrm{curl} \circ \mathrm{curl}$, which will be mentioned later on.

How Sobolev spaces of different regularities, as well as Sobolev spaces and other function spaces can be related to each other is made clear by the following three theorems.

Theorem 2.15 (Sobolev Embedding Theorem, [8, Theorem 2.5.4]) *Let $\Omega \subseteq \mathbb{R}^d$ be a bounded C^k-domain. Let m be an integer such that $m < \ell - d/2$ holds for some real number $\ell < k$. Then the embedding $H^\ell(\Omega) \hookrightarrow C^m(\Omega)$ is well-defined and continuous. Specifically, for $m = 0$ this reduces to the continuous embedding $H^{d/2+\epsilon} \hookrightarrow C^0(\Omega)$ for all $\epsilon > 0$.*

Theorem 2.16 (Inclusion of Sobolev Spaces, [8, Proposition 2.5.2]) *Let $s \leq t$. Then the embedding $H^t(\Omega) \hookrightarrow H^s(\Omega)$ is well-defined and continuous.*

In the sense of the Theorem 2.16, we will commonly write $H^t(\Omega) \subseteq H^s(\Omega)$.

Theorem 2.17 (Interpolation Inequality, [9, Definition 2.4.1, Theorem 4.1.2]) *Let $0 \leq s_1 \leq s_2$ and $0 \leq t_1 \leq t_2$ be integers and let Ω be a bounded Lipschitz domain. For $\sigma \in [0, 1]$, if $T: H^{s_j}(\Omega) \to H^{t_j}(\Omega)$ is a bounded linear operator for both $j = 1, 2$, with*

$$\|Tu\|_{H^{t_j}(\Omega)} \leq C_j \|u\|_{H^{s_j}(\Omega)},$$

for two constants C_1 and C_2, then we find

$$\|Tu\|_{H^{(1-\sigma)\cdot t_1+\sigma\cdot t_2}(\Omega)} \leq C_1^{1-\sigma} C_2^{\sigma} \|u\|_{H^{(1-\sigma)\cdot s_1+\sigma\cdot s_2}(\Omega)}.$$

The results of the last three theorems carry over to Sobolev spaces defined via graph norm, where a proof of the interpolation inequality for graph norm spaces is discussed e.g. in [10, Proposition 4.14].

2.4 Tensor Products of Sobolev Spaces

Within this thesis we will heavily rely on tensor product structures and will thus also encounter tensor products of function spaces. In this section, we will compile some elementary results and definitions. The account will follow [11, Sect. II].

Definition 2.18 (*Simple Tensors of Functions,* [11, Eq. (II.1.5)]) We define a *simple tensor* $f \in H^s(0, 1) \otimes H^t(0, 1)$ as a function

$$f = f_1 \otimes f_2 := f_1 f_2$$

for $f_1 \in H^s(0, 1)$ and $f_2 \in H^t(0, 1)$.

Scalar products of tensor product Sobolev spaces are induced by their definition on simple tensors

$$\langle f_1 \otimes f_2, g_1 \otimes g_2 \rangle_{H^s(0,1)\otimes H^t(0,1)} = \langle f_1, g_1 \rangle_{H^s(0,1)} \langle f_2, g_2 \rangle_{H^t(0,1)},$$

inducing the norm

$$\|f_1 \otimes f_2\|_{H^s(0,1)\otimes H^t(0,1)} = \|f_1\|_{H^s(0,1)} \|f_2\|_{H^t(0,1)}.$$

Tensor product Sobolev spaces can then be defined through completion of finite linear combinations of simple tensors w.r.t. this norm.

Definition 2.19 (*Tensor Product Sobolev Spaces,* [12, Sect. II.4]) For integers $s, t \geq 0$ the tensor product space $H^s(0,1) \otimes H^t(0,1)$ is defined as the closure of

$$\left\{ \sum_{0 \leq i < N_1} \sum_{0 \leq j < N_2} c_{j,i} f_1 f_2 \colon c_{j,i} \in \mathbb{C},\ f_1 \in H^s(0,1),\ f_2 \in H^t(0,1) \right\}$$

w.r.t. $\|\cdot\|_{H^s(0,1) \otimes H^t(0,1)}$.

The following alternative characterisation will be useful when reasoning about these spaces.

Theorem 2.20 (Alternative Characterisation, [12, Theorem II.10]) *For natural numbers $s, t \geq 0$, the identification*

$$H^s(0,1) \otimes H^t(0,1) = \{ f \in L^2(\square) \colon \partial_x^\alpha \partial_y^\beta f \in L^2(\square) \text{ for all } \alpha \leq s, \beta \leq t \} \tag{2.4}$$

holds.

This allows us to provide some elementary inclusions.

Corollary 2.21 (Inclusions of Tensor Product Sobolev Spaces) *For integers s it holds that*

$$H^s(\square) \supsetneq H^s(0,1) \otimes H^s(0,1) \supsetneq H^{2s}(\square), \quad s > 0, \tag{2.5}$$

where equality of these spaces holds only for $s = 0$. Moreover, it holds that

$$\big(H^1(0,1) \otimes L^2(0,1)\big) \times \big(L^2(0,1) \otimes H^1(0,1)\big) \subseteq \boldsymbol{H}^0(\mathrm{div}, \square). \tag{2.6}$$

Proof The first assertion follows directly when comparing (2.4) to the representation

$$H^s(\square) := \{ f \in L^2(\square) \colon \partial_x^\alpha \partial_y^\beta f \in L^2(\square) \text{ for all } \alpha + \beta \leq s \}.$$

For $\boldsymbol{f} = (f_x, f_y)^\intercal$, the second inclusion follows immediately from the inequality

$$|\partial_x f_x| + |\partial_y f_y| \geq |\partial_x f_x + \partial_y f_y| = |\operatorname{div}(\boldsymbol{f})|,$$

which estimates the seminorm terms of both spaces in (2.6). ■

It only remains to define the action of tensor products of linear operators through density of finite linear combinations of simple tensors.

Definition 2.22 (*Tensor Products of Linear Operators,* [13, Definition 1.29]) Let X and Y denote either Sobolev spaces on $(0,1)$ or $\mathbb{C}$. Tensor products of linear operators $T_A \colon H^s(0,1) \to X$ and $T_B \colon H^t(0,1) \to Y$ are defined for finite linear combinations of simple tensors via

$$(T_A \otimes T_B)\left(\sum_{0\le i<N_1}\sum_{0\le j<N_2} c_{j,i} f_1 f_2\right) = \sum_{0\le i<N_1}\sum_{0\le j<N_2} c_{j,i}(T_A(f_1))(T_B(f_2))$$

and on $H^s(0,1) \otimes H^t(0,1)$ through the density relation stated in Definition 2.19.

2.5 Sobolev Spaces on Boundaries

We are interested in function spaces on compact boundaries $\Gamma = \partial\Omega$ of Lipschitz domains Ω. As commonly done, we define the corresponding spaces on Γ via charts and partitions of unity. In the following, we will briefly go through this process, since an intuition about it is helpful for arguments to come.

Following the notation of the previous definition, let $\Gamma_j = \Gamma \cap W_j$ be invoked by a Lipschitz mapping θ and let $\widehat{\Gamma}_j \subseteq \mathbb{R}^{d-1}$ be given as the set

$$\widehat{\Gamma}_j := \{\boldsymbol{x} : (\boldsymbol{x}, \theta(\boldsymbol{x})) \in W_j \cap \Gamma\}.$$

Now, for a Lebesgue-measurable function f one can define integration over Γ_j via

$$\int_{\Gamma_j} f(\boldsymbol{x})\, \mathrm{d}\Gamma_{\boldsymbol{x}} := \int_{\widehat{\Gamma}_j} f(\boldsymbol{x}', \theta(\boldsymbol{x}'))\sqrt{\det(\boldsymbol{d}\theta(\boldsymbol{x}')^\intercal \boldsymbol{d}\theta(\boldsymbol{x}'))}\, \mathrm{d}\boldsymbol{x}', \tag{2.7}$$

where $\boldsymbol{d}\theta(\boldsymbol{x}')$ denotes the Jacobian of θ at a point $\boldsymbol{x}' \in \widehat{\Gamma}$. This enables us to give meaning to the spaces $H^s(\Gamma_j)$, as well as their norms and scalar products, in analogy to domains, by using (2.7) as a means of integration. Thus, for $u \in L^2(\Gamma)$ and hypographs Γ_j induced by a mapping θ we can set

$$u_\theta(\boldsymbol{x}') = u(\boldsymbol{x}', \theta(\boldsymbol{x}'))$$

for $\boldsymbol{x}' \in \mathbb{R}^{d-1}$. This allows us to define the space $H^s(\Gamma_j)$ as

$$H^s(\Gamma_j) = \left\{u \in L^2(\Gamma_j) : u_\theta \in H^s\big(\widehat{\Gamma}_j\big)\right\},$$

given that θ is regular enough. This space can then be equipped with the scalar product

$$\langle u, v\rangle_{H^s(\Gamma_j)} = \langle u_\theta, v_\theta\rangle_{H^s(\widehat{\Gamma}_j)}.$$

We can now define the following.

Definition 2.23 (*Sobolev Spaces on Manifolds,* [1, p. 99]) Let Γ be a Lipschitz boundary and let $s > 0$. Let $\{\phi_i\}_{i\in J}$ be a partition of unity subordinate to the finite open cover $\{W_i\}_{i\in I}$ of Γ, i.e., a family of $C^\infty(W_j)$-functions such that $\sum_{i\in J}\phi_j(x) = 1$ for all $x \in \Gamma$. Then, one can define an inner product on Γ via

$$\langle u, v\rangle_{H^s(\Gamma)} = \sum_{i\in J} \langle \phi_j u, \phi_j v\rangle_{H^s(\Gamma_j)},$$

where, as above, $\Gamma_j = \Gamma \cap W_j$. The corresponding spaces $H^s(\Gamma)$ and, analogously the vector-valued spaces $\boldsymbol{H}^s(\Gamma)$, can now be defined be via the induced norms, cf. (2.1).

Remark 2.24 ([1, p. 98f.]) Two remarks on these function spaces on Γ are due.

1. A different choice of partition $(\Gamma_j, W_j)_{j\in J}$ yields equivalent scalar products and norms when applying Definition 2.23. Moreover, (2.7) makes it clear that the range of s is limited by the smoothness of Γ: If the Lipschitz mappings $\{\theta_j\}_{j\in J}$ can be chosen to be $C^{k-1,1}$ for $k \geq 0$, then s can be chosen as $s \in [-k, k]$.
2. The assertions of Theorems 2.15, 2.16, and 2.17 carry over to the spaces on manifolds.

Given Definition 2.23, we define $H^{-s}(\Gamma)$ to denote the dual space to $H^s(\Gamma)$ in the sense of Definition 2.3, with $L^2(\Gamma)$ as a pivot space. The vector-valued spaces are treated analogously.

These spaces on boundaries relate to spaces on the domain through the restriction of functions in Ω onto Γ. Formally, this can be realised through the so-called *trace operators*.

Note that since θ is Lipschitz, one can show that the Jacobian $\boldsymbol{d}\theta$ exists almost everywhere, cf. [1, p. 96.]. This immediately ensures the existence of the *outward unit normal vector* $\boldsymbol{n}_{\boldsymbol{x}}$ for almost all $\boldsymbol{x} \in \Gamma$, as well.

Definition 2.25 (*Trace Operators,* [3, 8]) Let $u\colon \mathbb{R}^3 \to \mathbb{C}$ and $\boldsymbol{u}\colon \mathbb{R}^3 \to \mathbb{C}^3$ be smooth. For almost all $\boldsymbol{x}_0 \in \Gamma$ we can define the

Dirichlet trace	$\gamma_0(u)(\boldsymbol{x}_0) := \lim\limits_{\boldsymbol{x}\to\boldsymbol{x}_0} u(\boldsymbol{x}),$
tangential trace	$\boldsymbol{\gamma}_0(\boldsymbol{u})(\boldsymbol{x}_0) := \lim\limits_{\boldsymbol{x}\to\boldsymbol{x}_0} \boldsymbol{u}(\boldsymbol{x}) - \boldsymbol{n}_{\boldsymbol{x}_0}(\boldsymbol{u}(\boldsymbol{x})\cdot\boldsymbol{n}_{\boldsymbol{x}_0}),$
rotated tangential trace	$\boldsymbol{\gamma}_t(\boldsymbol{u})(\boldsymbol{x}_0) := \lim\limits_{\boldsymbol{x}\to\boldsymbol{x}_0} \boldsymbol{u}(\boldsymbol{x}) \times \boldsymbol{n}_{\boldsymbol{x}_0},$
and *normal trace*	$\gamma_n(\boldsymbol{u})(\boldsymbol{x}_0) := \lim\limits_{\boldsymbol{x}\to\boldsymbol{x}_0} \boldsymbol{u}(\boldsymbol{x}) \cdot \boldsymbol{n}_{\boldsymbol{x}_0}.$

Equipped with the superscript ^{int} we will refer to them *interior traces* and choose $\boldsymbol{x} \in \Omega$. For the superscript ^{ext}, they will be referred to as *exterior traces*, where we will set $\boldsymbol{x} \in \Omega^c$. When the choice of interior or exterior is clear from the context, i.e., when the domain of the trace is clearly stated, we will omit the superscript for ease of notation.

The trace operators are visualised in Fig. 2.1.

We will now spend the remainder of this section with a discussion of the ranges of said trace operators. For this we define the trace space

$$\boldsymbol{H}^s_\times(\Gamma) := \boldsymbol{\gamma}_t(\boldsymbol{H}^{s+1/2}(\Omega)), \quad \text{for} \quad 0 < s < 1,$$

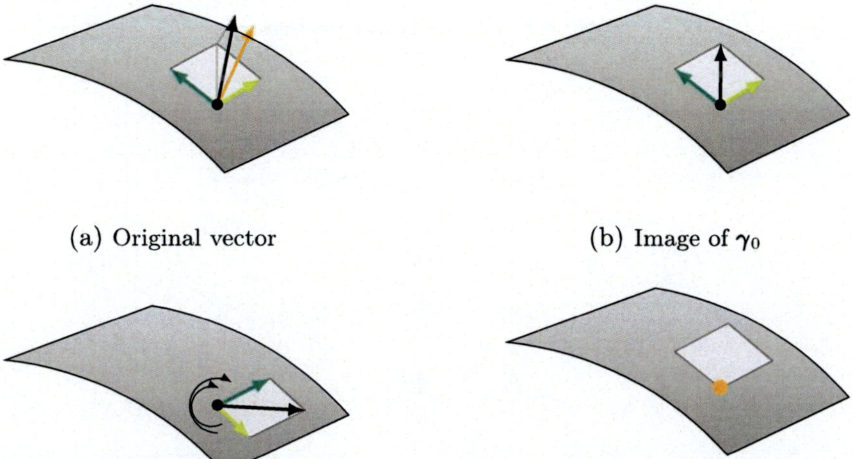

(a) Original vector

(b) Image of γ_0

(c) Image of γ_t

(d) Image of γ_n

Fig. 2.1 Visualisation of the trace operators. The black arrow corresponds to a pointwise value of a vector field, which can be decomposed into two tangential components (green and yellow), as well as a normal component (orange)

where the subscript $\times$ will be used to denote the rotation around the normal.

The space $\boldsymbol{H}_\times^{-s}(\Gamma)$ denotes the corresponding dual space w.r.t. the duality pairing

$$\langle \boldsymbol{\nu}, \boldsymbol{\mu} \rangle_\times := \int_\Gamma (\boldsymbol{\nu} \times \boldsymbol{n}) \cdot \boldsymbol{\mu} \, \mathrm{d}\Gamma, \tag{2.8}$$

cf. [3, Definition 1]. Note that $\boldsymbol{H}_\times^s(\Gamma)$ might not coincide with $\boldsymbol{H}^s(\Gamma)$ understood in a componentwise sense, since jumps of the normal might break existing continuities in the preimage. However, this identification holds for smooth-manifolds, cf. [14].

Defining

$$\boldsymbol{H}_\times^{-1/2}(\mathrm{div}_\Gamma, \Gamma) := \gamma_t(\boldsymbol{H}^0(\mathbf{curl}, \Omega)),$$

together with its rotated counterpart

$$\boldsymbol{H}^{-1/2}(\mathrm{curl}_\Gamma, \Gamma) := \gamma_0(\boldsymbol{H}^0(\mathbf{curl}, \Omega)),$$

one can use density arguments to extend the trace operators to a weak setting with domain Ω or Ω^c, respectively. We summarise the resulting mapping properties, as presented by McLean [1, Theorem 3.37], and Buffa and Hiptmair [3, Theorems 1, 3].

Theorem 2.26 (Mapping Properties of the Trace Operators, [1, 3]) *For the trace operators, the following properties hold.*

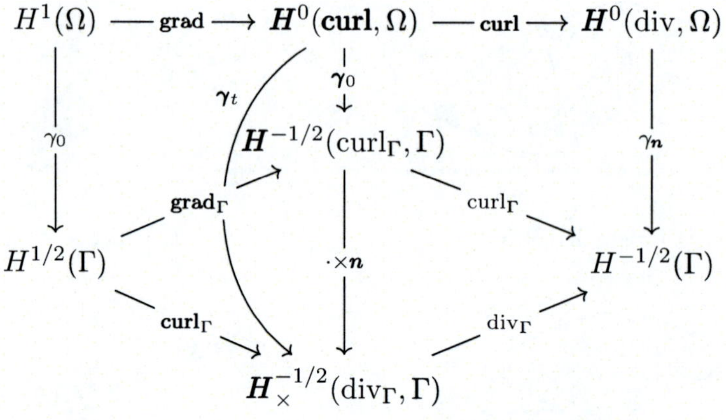

Fig. 2.2 Two dimensional de Rham complex on the boundary, induced by application of the trace operators to the three-dimensional complex in the domain

1. *The trace operator* $\gamma_0 \colon H^s(\Omega) \to H^{s-1/2}(\Gamma)$ *is linear, continuous and surjective, with a continuous right inverse for* $1/2 < s < 3/2$.
2. *The operator* $\boldsymbol{\gamma}_0 \colon \boldsymbol{H}^0(\mathbf{curl}, \Omega) \to \boldsymbol{H}^{-1/2}(\mathrm{curl}_\Gamma, \Gamma)$ *is linear, continuous, surjective, and possesses a continuous right inverse.*
3. *The operator* $\boldsymbol{\gamma}_t \colon \boldsymbol{H}^0(\mathbf{curl}, \Omega) \to \boldsymbol{H}_\times^{-1/2}(\mathrm{div}_\Gamma, \Gamma)$ *is linear, continuous, surjective, and possesses a continuous right inverse.*
4. *The operator* $\gamma_n \colon \boldsymbol{H}^0(\mathrm{div}, \Omega) \to H^{-1/2}(\Gamma)$ *is linear, continuous and surjective.*

The resulting connectivity between functions spaces in the domain and on the boundary is depicted in Fig. 2.2. Note that the Hilbert spaces $\boldsymbol{H}^{-1/2}(\mathrm{curl}_\Gamma, \Gamma)$ and $\boldsymbol{H}_\times^{-1/2}(\mathrm{div}_\Gamma, \Gamma)$ can be defined in terms of spaces induced by a graph norm, in analogy to their counterparts in the volume. To do so, using the surjectivity of the traces as well as the existence and continuity of right-inverse operators, we can introduce the following.

Definition 2.27 (*Surface Differential Operators*) Given Definition 2.25, we define the surface differential operators $\mathbf{grad}_\Gamma$, $\mathbf{curl}_\Gamma$, curl_Γ, and div_Γ such that the diagram in Fig. 2.2 commutes.

This yields a definition of spaces with graph norm on Γ, where $\boldsymbol{H}^s(\mathrm{div}_\Gamma, \Gamma)$ will be of specific importance. Specifically, for the traces of $\boldsymbol{H}(\mathbf{curl}, \Omega)$ we can proceed as in [10, p. 8] and characterise

$$\boldsymbol{H}_\times^{-1/2}(\mathrm{div}_\Gamma, \Gamma) = \{\boldsymbol{f} \in \boldsymbol{H}_\times^{-1/2}(\Gamma) \colon \mathrm{div}_\Gamma \boldsymbol{f} \in H^{-1/2}(\Gamma)\}. \tag{2.9}$$

2.6 A Model of Electromagnetism

We will now introduce the physical model for the problems to be investigated. This section is largely based on the books of Jackson [15]and Monk [5].

Let T denote a (possibly infinite) time interval in $\mathbb{R}$. Furthermore, let $\Omega \subseteq \mathbb{R}^3$ denote a domain, and let ∂_t denote derivation w.r.t. time. Via vector valued functions

$$\mathcal{E}\colon \Omega \times T \to \mathbb{R}^3, \quad \text{called } \textit{electric field (intensity)}, \tag{2.10}$$

$$\mathcal{H}\colon \Omega \times T \to \mathbb{R}^3, \quad \text{called } \textit{magnetic field (intensity)}, \tag{2.11}$$

$$\mathcal{D}\colon \Omega \times T \to \mathbb{R}^3, \quad \text{called } \textit{electric displacement}, \tag{2.12}$$

$$\mathcal{B}\colon \Omega \times T \to \mathbb{R}^3, \quad \text{called } \textit{magnetic induction}, \tag{2.13}$$

$$\mathcal{J}\colon \Omega \times T \to \mathbb{R}^3, \quad \text{called } \textit{current density}, \tag{2.14}$$

and the scalar function

$$\varrho\colon \Omega \times T \to \mathbb{R}, \quad \text{called } \textit{charge density}, \tag{2.15}$$

one can model electromagnetic fields by the well-known *Maxwell's Equations*

$$\mathbf{curl}\,\mathcal{E} = -\partial_t \mathcal{B}, \quad \text{called } \textit{Faraday's law}, \tag{2.16}$$

$$\mathbf{curl}\,\mathcal{H} = \partial_t \mathcal{D} + \mathcal{J}, \quad \text{called } \textit{Ampère'slaw}, \tag{2.17}$$

$$\operatorname{div} \mathcal{D} = \varrho, \quad \text{called } \textit{Gauss's law}, \text{ and} \tag{2.18}$$

$$\operatorname{div} \mathcal{B} = 0, \quad \text{referred to as } \textit{Gauss's magnetism law}. \tag{2.19}$$

These quantities can be equipped with their corresponding SI units, see e.g. the book of Jackson [15, Sect. 1.2]. For reasons of notational simplicity, and since most arguments in this thesis will be independent of units, we will not refer to them.

For suitable vector fields $\boldsymbol{E}, \boldsymbol{H}, \boldsymbol{D}, \boldsymbol{B}, \boldsymbol{J} : \Omega \to \mathbb{C}^3$ as well as a complex-valued scalar quantity $\widehat{\varrho}\colon \Omega \to \mathbb{C}$ we assume sinusoidal oscillation of currents and charges, i.e., relations of the form

$$\begin{aligned} \mathcal{E} &= \mathfrak{Re}\big(e^{-i\omega t}\boldsymbol{E}\big), & \mathcal{H} &= \mathfrak{Re}\big(e^{-i\omega t}\boldsymbol{H}\big), \\ \mathcal{D} &= \mathfrak{Re}\big(e^{-i\omega t}\boldsymbol{D}\big), & \mathcal{B} &= \mathfrak{Re}\big(e^{-i\omega t}\boldsymbol{B}\big), \\ \mathcal{J} &= \mathfrak{Re}\big(e^{-i\omega t}\boldsymbol{J}\big), & \varrho &= \mathfrak{Re}\big(e^{-i\omega t}\widehat{\varrho}\big). \end{aligned}$$

Under these assumptions the system (2.16)–(2.19) can be reduced to the so-called *time-harmonic* system of equations

$$\mathbf{curl}\,\boldsymbol{E} = i\omega \boldsymbol{B}, \tag{2.20}$$

$$\mathbf{curl}\,\boldsymbol{H} = -i\omega \boldsymbol{D} + \boldsymbol{J}, \tag{2.21}$$

$$\operatorname{div} \boldsymbol{D} = \widehat{\varrho}, \tag{2.22}$$

$$\operatorname{div} \boldsymbol{B} = 0, \tag{2.23}$$

cf. Sect. 1.2 in the book of Monk [5].

The electric displacement and the magnetic induction can be expressed in terms of the electric and magnetic field via material parameters.

The material parameter *electric permittivity* and *magnetic permeability* denoted by ε and μ, respectively, will be assumed to be scalar constants. This modelling choice yields *material laws* of the form

$$\boldsymbol{D} = \varepsilon \boldsymbol{E} \quad \text{and} \tag{2.24}$$

$$\boldsymbol{B} = \mu \boldsymbol{H}. \tag{2.25}$$

Remark 2.28 (*Differential Forms*) The framework of differential forms links naturally to the structure of electromagnetism. In this sense, the volumetric section of Fig. 2.2 relates to 0-, 1-, and 2-forms, where the classical endpoint of this exact sequence, $L^2(\Omega)$, can be identified with the 3-forms. In this sense the electromagnetic quantities can be identified with the function spaces of the De Rham sequence with $\boldsymbol{E}, \boldsymbol{H} \in \boldsymbol{H}(\mathbf{curl}, \Omega)$, and $\boldsymbol{B}, \boldsymbol{D} \in \boldsymbol{H}(\operatorname{div}, \Omega)$. The material laws manifest as applications of the Hodge operator, showcasing, together with Faraday's and Ampère's law, the isomorphism of 1- and 2-forms in three dimensions.

To keep the upcoming analysis palatable, a further discussion of this has to be put out of the scope of this document. More about this can be found in the works of Bossavit [16], while an account specific to Maxwell BEM can be found in the work of Kurz and Auchmann [17].

Assuming the absence of currents, i.e., $\boldsymbol{J} = 0$, one can insert (2.25) in (2.20) and (2.24) in (2.21), arriving at

$$\mathbf{curl} \frac{1}{i\omega\mu} \mathbf{curl} \boldsymbol{E} + i\omega\varepsilon \boldsymbol{E} = 0. \tag{2.26}$$

Due to linearity of the **curl** operator, we can move the material parameters together and define the non-negative wavenumber $\kappa = \omega\sqrt{\varepsilon\mu}$. This means, that any vector field $\boldsymbol{E}$ which fulfils the equation

$$\mathbf{curl}\mathbf{curl} \boldsymbol{E} - \kappa^2 \boldsymbol{E} = 0, \qquad \kappa > 0, \tag{2.27}$$

is a valid model of an electric field in the given setting and in the absence of currents.

The electric wave equation will be the governing partial differential equation throughout the thesis. However, formulating this equation in terms of a well-posed boundary value problem requires the introduction of boundary conditions.

We will model all geometries as perfect electric conductors, resulting in the application of the so-called *perfect electric conducting (PEC)* boundary condition, which can be derived from the absence of a field within mediums of infinite conductivity.

Since this boundary condition will be applied to all problems discussed within this thesis, we will state it as an assumption.

Assumption 2.29 (*Perfect Electric Conductor,* [5, Eq. (1.17)]) For any solution to (2.27) we assume that

$$\gamma_t \boldsymbol{E} = \boldsymbol{E} \times \boldsymbol{n} = 0 \quad \text{on } \Gamma. \tag{2.28}$$

Note that, through Maxwell's Equations, as well as the material relations, the PEC boundary condition implies $\boldsymbol{H} \cdot \boldsymbol{n} = 0$ on Γ.

When interested in exterior problems, we must also apply a condition at infinity. This is given by the so-called *Silver–Müller Radiation Condition.*

Assumption 2.30 (*Silver–Müller Radiation Condition,* [17, p. 29]) For $\boldsymbol{x} \in \Omega^c$ and any solution $\boldsymbol{E}$ to (2.27) we assume that

$$\left| \mathbf{curl}\ \boldsymbol{E}(\boldsymbol{x}) \times \frac{\boldsymbol{x}}{|\boldsymbol{x}|} - i\kappa \boldsymbol{E}(\boldsymbol{x}) \right| = o\left(\frac{1}{|\boldsymbol{x}|}\right), \qquad |\boldsymbol{x}| \to \infty.$$

We remark that the radiation condition as formulated above is more specific than the ones often cited, see e.g. [5, Eq. (1.22)] or [10, Eq. (55)], which only demand that the limit vanishes. When formulated as above the physical interpretation becomes clear: For increasing distance to the scatterer, the scattered field should start to resemble that of a spherical wave going out from the scatterer. This means that for a large distance to the scatterer, geometric details become negligible within the observed field.

The assumptions above enable us to formulate the two problems central to this thesis.

Problem 2.31 (*Scattering Problem,* [5, Sect. 1.4.1]) Given a Lipschitz domain Ω with boundary Γ, a wavenumber κ, and incident wave $\boldsymbol{E}_i \in \boldsymbol{H}^0_{\mathrm{loc}}(\mathbf{curl}, \mathbb{R}^3)$, we are looking for the scattered field $\boldsymbol{E}_s \in \boldsymbol{H}_{\mathrm{loc}}(\mathbf{curl}^2, \Omega^c)$ which satisfies

$$\begin{aligned} \mathbf{curl}\mathbf{curl}\boldsymbol{E}_s - \kappa^2 \boldsymbol{E}_s &= 0, && \text{in } \Omega^c, \\ \gamma_t \boldsymbol{E}_s &= -\gamma_t \boldsymbol{E}_i, && \text{on } \Gamma, \\ \left| \mathbf{curl}\ \boldsymbol{E}_s(\boldsymbol{x}) \times \frac{\boldsymbol{x}}{|\boldsymbol{x}|} - i\kappa \boldsymbol{E}_s(\boldsymbol{x}) \right| &= o\left(\frac{1}{|\boldsymbol{x}|}\right), && |\boldsymbol{x}| \to \infty. \end{aligned}$$

Note that the second equation of Problem 2.31 ensures that Assumption 2.29 is satisfied by the total field $\boldsymbol{E} = \boldsymbol{E}_s + \boldsymbol{E}_i$.

Moreover, an analogue of Problem 2.31 can be formulated for the interior domain Ω, where the radiation condition can be simply omitted.

Another application of the techniques mentioned within this work is the solution of eigenvalue problem induced by (2.27), i.e., the approximation of resonant frequencies and corresponding electromagnetic fields within a resonator.

Problem 2.32 (*Resonance Problem,* [5, Sect. 1.4.2]) Given a scatterer Ω with Lipschitz boundary Γ, we are looking for an eigenpair $(\boldsymbol{E}, \kappa^2) \in \boldsymbol{H}(\mathbf{curl}^2, \Omega) \times \mathbb{R}_+$ with $\boldsymbol{E} \neq 0$, $\kappa \neq 0$ that satisfies

$$\begin{aligned} \mathbf{curl}\mathbf{curl}\boldsymbol{E} - \kappa^2 \boldsymbol{E} &= 0, && \text{in } \Omega, \\ \boldsymbol{E} \times \boldsymbol{n} &= 0, && \text{on } \Gamma. \end{aligned}$$

For the remainder of this thesis we will call a wavenumber κ *non-resonant* if there does not exist an $\boldsymbol{E}$ such that $(\boldsymbol{E}, \kappa)$ is a solution to Problem 2.32.

2.7 Representation of Fields via Boundary Integral Equations

For the remainder of the chapter we will focus on Problem 2.31. Indeed, any solution to this problem can be derived via a boundary integral formulation.

Although its derivation is technically involved, it is not instructive for the topics discussed within this thesis. Thus, we will only state the assertion. The derivations reviewed here are originally due to Stratton and Chu, see e.g. [18], but have also been discussed in a plethora of other works, see e.g. [3, 19]. The version reviewed here resembles the one presented by Buffa and Hiptmair [3, Theorem 6].

For any wavenumber $\kappa \in \mathbb{C}$ let $u_\kappa^*(\boldsymbol{x}, \boldsymbol{y}) \colon \mathbb{R}^3 \times \mathbb{R}^3 \to \mathbb{C}$ denote the *Helmholtz fundamental solution*

$$u_\kappa^*(\boldsymbol{x}, \boldsymbol{y}) := \frac{e^{-i\kappa|\boldsymbol{x}-\boldsymbol{y}|}}{4\pi|\boldsymbol{x}-\boldsymbol{y}|}. \tag{2.29}$$

For notational convenience, we define the *Maxwell double layer potential*

$$\mathcal{W}_\kappa \colon \boldsymbol{H}_\times^{-1/2}(\mathrm{div}_\Gamma, \Gamma) \to \boldsymbol{H}_{\mathrm{loc}}(\mathbf{curl}^2, \Omega \cup \Omega^c)$$

as

$$\mathcal{W}_\kappa(\boldsymbol{\mu})(\boldsymbol{x}) = \mathbf{curl}_x \int_\Gamma u_\kappa^*(\boldsymbol{x}, \boldsymbol{y}) \boldsymbol{\mu}(\boldsymbol{y}) \,\mathrm{d}\Gamma_{\boldsymbol{y}},$$

and the *Maxwell single layer potential*

$$\mathcal{V}_\kappa \colon \boldsymbol{H}_\times^{-1/2}(\mathrm{div}_\Gamma, \Gamma) \to \boldsymbol{H}_{\mathrm{loc}}(\mathbf{curl}^2, \Omega \cup \Omega^c)$$

as

$$\mathcal{V}_\kappa(\boldsymbol{\mu})(\boldsymbol{x}) = \int_\Gamma u_\kappa^*(\boldsymbol{x}, \boldsymbol{y}) \boldsymbol{\mu}(\boldsymbol{y}) \,\mathrm{d}\Gamma_{\boldsymbol{y}} + \frac{1}{\kappa^2} \mathbf{grad}_x \int_\Gamma u_\kappa^*(\boldsymbol{x}, \boldsymbol{y}) (\mathrm{div}_\Gamma \boldsymbol{\mu})(\boldsymbol{y}) \,\mathrm{d}\Gamma_{\boldsymbol{y}}$$

for sufficiently smooth functions μ. We extend their meaning to the weak setting by density arguments. The resulting mapping properties are reviewed by Buffa and Hiptmair [3, Theorem 5], who also show that the image of both potentials is divergence-free.

Armed with these operators we can now introduce the well-known boundary integral formulation.

Theorem 2.33 (Stratton–Chu Representation Formula, [3, Theorem 6]) *Any* $\boldsymbol{E}_s \in \boldsymbol{H}^0_{\mathrm{loc}}(\mathbf{curl}^2, \Omega^c)$ *that is the solution to Problem 2.31 can be represented by*

$$\boldsymbol{E}_s = -\mathcal{W}_\kappa\big(\gamma_t^{\mathrm{ext}}(\boldsymbol{E}_s)\big) - \mathcal{V}_\kappa\big((\gamma_t^{\mathrm{ext}} \circ \mathbf{curl})(\boldsymbol{E}_s)\big).$$

Moreover, any $\boldsymbol{E}_s \in \boldsymbol{H}^0(\mathbf{curl}^2, \Omega)$ *which solves the corresponding interior problem within* Ω *can be represented as*

$$\boldsymbol{E}_s = \mathcal{W}_\kappa\big(\gamma_t^{\mathrm{int}}(\boldsymbol{E}_s)\big) + \mathcal{V}_\kappa\big((\gamma_t^{\mathrm{int}} \circ \mathbf{curl})(\boldsymbol{E}_s)\big).$$

Remark 2.34 (*Null-Property,* [17, p. 33]) The evaluation of the interior representation in the exterior domain yields 0, and vice versa. This is sometimes referred to as *null-property of the representation formula.*

Remark 2.35 (*Cauchy Data,* [10, Definition 3]) The Stratton–Chu formula already allows for a solution of the boundary value problem given $\gamma_t(\boldsymbol{E}_s)$, which is determined through the incoming wave because $\boldsymbol{E}_s + \boldsymbol{E}_i$ must fulfill Assumption 2.29. In this case $\gamma_t \circ \mathbf{curl}\boldsymbol{E}_s$ is the unknown, and the pair $(\gamma_t\boldsymbol{E}_s, \gamma_t \circ \mathbf{curl}\boldsymbol{E}_s)$ is called *Cauchy data.*

Let $\boldsymbol{E}^{\mathrm{ext}}$ denote the solution to an exterior scattering problem induced by the incoming wave $\boldsymbol{E}_i$, i.e., governed by the boundary condition

$$\gamma_t^{\mathrm{ext}}\boldsymbol{E}_s^{\mathrm{ext}} = -\gamma_t^{\mathrm{ext}}\boldsymbol{E}_i.$$

Then we can find the solution $\boldsymbol{E}^{\mathrm{int}}$ to the corresponding interior problem, with boundary data generated by

$$\gamma_t^{\mathrm{ext}}\boldsymbol{E}_s^{\mathrm{ext}} = \gamma_t^{\mathrm{int}}\boldsymbol{E}_s^{\mathrm{int}}.$$

Applying this to Theorem 2.33 and setting $\boldsymbol{w} := (\gamma_t^{\mathrm{ext}} - \gamma_t^{\mathrm{int}})(\mathbf{curl}\boldsymbol{E})$, together with the null-property yields the following.

Corollary 2.36 (Single-Layer Representation, [3, Sect. 7.2]) *If* $\boldsymbol{E}_s \in \boldsymbol{H}_{\mathrm{loc}}(\mathbf{curl}^2, \Omega \cup \Omega^c)$ *is a solution to the electric wave equation (2.27) on* $\Omega \cup \Omega^c$ *that also fulfills Assumptions 2.29 and 2.30, then it can be represented via*

$$\boldsymbol{E}_s = -\mathcal{V}_\kappa(\boldsymbol{w}).$$

Denoting the average of the interior and exterior rotated tangential trace by

$$\{\gamma_t\} = \frac{1}{2}(\gamma_t^{\mathrm{int}} + \gamma_t^{\mathrm{ext}}),$$

we define the *Maxwell single layer operator*

$$\mathbf{V}_\kappa := \{\gamma_t\} \circ \mathcal{V}_\kappa.$$

The Maxwell single layer operator allows us to recast the identity in Corollary 2.36 in terms of a variational problem on the boundary w.r.t. the duality pairing (2.8). This identity is famously known as the *electric field integral equation (EFIE)*.

Problem 2.37 (*Electric Field Integral Equation,* [10]) Let $\kappa > 0$ be a non-resonant wavenumber, and $\boldsymbol{E}_i$ an incident wave. Find a function $\boldsymbol{w} \in \boldsymbol{H}_\times^{-1/2}(\mathrm{div}_\Gamma, \Gamma)$ such that

$$\langle \mathbf{V}_\kappa \boldsymbol{w}, \boldsymbol{\mu} \rangle_\times = -\langle \gamma_t \boldsymbol{E}_i, \boldsymbol{\mu} \rangle_\times$$

holds for all $\boldsymbol{\mu} \in \boldsymbol{H}_\times^{-1/2}(\mathrm{div}_\Gamma, \Gamma)$.

This unknown is a physical quantity known as surface current [15, p. 194], and throughout this thesis, we will use the electric field integral equation to represent the solution of scattering problems via Corollary 2.36.

Existence and uniqueness of the solution to the EFIE, and other variational problems that can be derived from Theorem 2.33 can be shown since they satisfy a so-called *generalised Gårding inequality*, see [3, Lemma 10] for non-resonant κ. The continuous problems have been discussed extensively in the literature, thus, this thesis will focus on their discrete counterparts which can be used to formalise numerical approximations to the continuous solution.

Remark 2.38 (*Representation and Solvability*) While solvability of Problem 2.37 can only be guaranteed for non-resonant wavenumbers, resonant fields can still be represented by means of Corollary 2.36, cf. [3].

The existence and uniqueness in the discrete sense depend, among other things, on the specific choice of a discrete space. For classical elements, for example, *Raviart–Thomas elements*, see the dissertation of Zaglmayr [20] for definitions, existence and uniqueness assertions have been provided, e.g. in the paper by Buffa and Christiansen [10], which is exclusively devoted to this task. The upcoming chapter establishes analogue assertions for the isogeometric setting.

References

1. McLean W (2000) Strongly elliptic systems and boundary integral equations. Cambridge University Press, Cambridge
2. Wloka J (1987) Partial differential equations. Cambridge University Press, Cambridge
3. Buffa A, Hiptmair R (2003) Galerkin boundary element methods for electromagnetic scattering. In: Ainsworth M, Davies P, Duncan D, Rynne B, Martin P (eds) Topics in computational wave propagation. Springer, Berlin, pp 83–124
4. Céa J (1964) Approximation variationnelle des problémes aux limites. Annales de l'Institut Fourier 14(2):345–444
5. Monk P (2003) Finite element methods for Maxwell's equations. Oxford University Press, Oxford
6. Babuška I (1969) Error-bounds for finite element method. J Numer Math 16(4):322–333
7. Xu J, Zikatanov L (2002) Some observations on Babuška and Brezzi theories. Numerische Mathematik 94(1):195–202
8. Sauter SA, Schwab C (2011) Boundary element methods. Springer, Berlin
9. Bergh J, Löfström J (1976) Interpolation spaces: an introduction. Grundlehren der mathematischen Wissenschaften. Springer, Berlin
10. Buffa A, Christiansen S (2003) The electric field integral equation on Lipschitz screens: definitions and numerical approximation. Numerische Mathematik 94(2):229–267
11. Dölz J (2017) Hierarchical matrix techniques for partial differential equations with random input data. Dissertation. Universitat Basel, Basel
12. Reed M, Simon B (1980) Methods of modern mathematical physics, volume 1: Functional analysis. Academic, San Diego
13. Light W, Cheney E (1985) Approximation theory in tensor product spaces, vol 1169, 1st edn. Lecture notes in mathematics. Springer, Berlin
14. Buffa A, Costabel M, Sheen D (2002) On traces for H(curl, Ω) in Lipschitz domains. J Math Anal Appl 276(2):845–867
15. Jackson JD (1998) Classical electrodynamics, 3rd edn. Wiley, New York
16. Bossavit A (1998) Computational electromagnetism: variational formulations, complementarity, edge elements. Academic, San Diego
17. Kurz S, Auchmann B (2012) Differential forms and boundary integral equations for Maxwell-type problems. In: Langer U, Schanz M, Steinbach O, Wendland WL (eds) Fast boundary element methods in engineering and industrial applications. Lecture notes in applied and computational mechanics, vol 63. Springer, Berlin, pp 1–62
18. Stratton JA (1941) Electromagnetic theory. IEEE Press, Piscataway
19. Xin J (2011) Boundary element approximation for Maxwell's eigenvalue problem. Dissertation. Karlsruhe Institute of Technology, Karlsruhe, Germany
20. Zaglmayr, S (2006) High order finite element methods for electromagnetic field computation. Dissertation. Johannes Kepler Universität, Linz, Austria

Chapter 3
Isogeometric Boundary Elements

We briefly review the basic notions of isogeometric methods. These notions create a framework in which pull-backs and push-forwards are used to conduct numerical computations without the introduction of a geometry error. Moreover, basis functions of higher regularity, i.e., splines, will be used as ansatz functions for this purpose.

We refer to works of Cottrell et al. [1] and Hughes et al. [2] for an introduction to isogeometric analysis and of Piegl and Tiller [3] and Schumaker [4] for more details on NURBS and spline theory. For notational convenience, we set $\square := (0, 1)^2$ throughout the entire thesis. The main results of this chapter have been first published in [5, 6], whose lines we follow for the introduction of fundamental concepts as well.

3.1 B-Spline Spaces

Of central importance is the notion of B-Splines, which are defined as follows.

Definition 3.1 (*B-Spline Basis,* [7, Sect. 2]) Let $\mathbb{K}$ be either $\mathbb{R}$ or $\mathbb{C}$ and p, k be integers with $0 \leq p < k$. We define a *p-open knot vector* Ξ as a set of the form

$$\Xi = \big\{ \underbrace{\xi_0 = \cdots = \xi_p}_{=0} < \xi_{p+1} \leq \cdots \leq \xi_{k-1} < \underbrace{\xi_k = \cdots = \xi_{k+p}}_{=1} \big\} \in [0, 1]^{k+p+1}.$$

The elements ξ_j are referred to as *knots* and we will assume their multiplicity to be at most p. We can then define the basis functions $\{b_i^p\}_{0 \leq i < k}$ on $[0, 1]$ for $p = 0$ as

© The Author(s), under exclusive license to Springer Nature Switzerland AG 2021
F. Wolf, *Analysis and Implementation of Isogeometric Boundary Elements for Electromagnetism*, Springer Theses,
https://doi.org/10.1007/978-3-030-61939-8_3

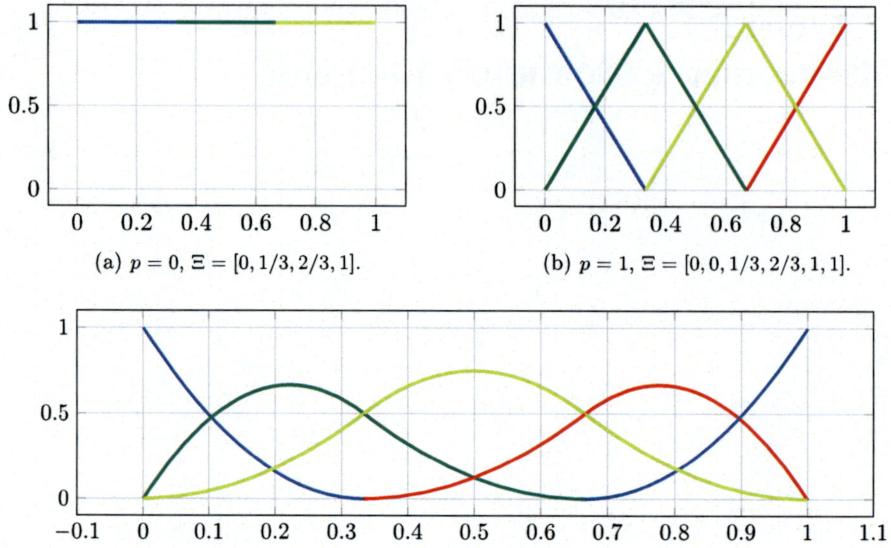

(a) $p = 0$, $\Xi = [0, 1/3, 2/3, 1]$.

(b) $p = 1$, $\Xi = [0, 0, 1/3, 2/3, 1, 1]$.

(c) $p = 2$, $\Xi = [0, 0, 0, 1/3, 2/3, 1, 1, 1]$.

Fig. 3.1 B-spline bases for $p = 0, 1, 2$ and open knot vectors with interior knots 1/3 and 2/3. Reprinted from [19], Copyright ©2018, with permission from Elsevier

$$b_i^0(x) = \begin{cases} 1, & \text{if } \xi_i \leq x < \xi_{i+1}, \\ 0, & \text{otherwise,} \end{cases}$$

and for $p > 0$ via the recursive relationship

$$b_i^p(x) = \frac{x - \xi_i}{\xi_{i+p} - \xi_i} b_i^{p-1}(x) + \frac{\xi_{i+p+1} - x}{\xi_{i+p+1} - \xi_{i+1}} b_{i+1}^{p-1}(x),$$

for all $0 \leq i < k$. The spline space $S_p(\Xi)$ is given as $\operatorname{span}(\{b_i^p\}_{0 \leq i < k})$. The integer k hereby denotes the dimension of the spline space. For a visualisation of B-splines see Fig. 3.1.

Remark 3.2 (*Derivative of a B-Spline,* [3]) By using the recursion formula, one can show through induction that the derivative of a B-spline of order p is a linear combination of two B-splines of order $p - 1$.

Let $d = 2, 3$, and let the knot vectors $\Xi_1, \ldots, \Xi_d$ be given. B-spline functions on the domain $[0, 1]^d$ are constructed through simple tensor product relationships for $p_{i_1,\ldots i_d} \in \mathbb{K}$ via

$$f(x_1, \ldots, x_d) := \sum_{0 \leq i_1 < k_1} \cdots \sum_{0 \leq i_d < k_d} p_{i_1,\ldots,i_d} \cdot b_{i_1}^{p_1}(x_1) \cdots b_{i_d}^{p_d}(x_d), \tag{3.1}$$

which allows *tensor product B-spline spaces*, denoted by

$$S_{p_1,\dots,p_d}(\Xi_1,\dots,\Xi_d)$$

to be defined. We will refer to non-empty intervals of the form $[\xi_i, \xi_{i+1}]$, $0 \le i < k$, and in the tensor product sense, non-empty sets of the form $[\xi_{i_1}, \xi_{i_1+1}] \times \cdots \times [\xi_{i_d}, \xi_{i_d+1}]$ as *elements* w.r.t. the knot vectors.

Definition 3.3 (*Locally Quasi-uniform Knot Vectors,* [7, Assumption 2.1]) For a p-open knot vector Ξ, let $h_i := \xi_{i+1} - \xi_i$. We define the *mesh size* h to be the maximal distance $h := \max_{p \le i < k} h_i$ between neighbouring knots. We call a knot vector *locally quasi-uniform* when for all elements $[\xi_{i_1}, \xi_{i_1+1}]$ and $[\xi_{i_2}, \xi_{i_2+1}]$ there exists a constant $\theta \ge 1$ such that the ratio $h_{i_1} \cdot h_{i_2}^{-1}$ satisfies $\theta^{-1} \le h_{i_1} \cdot h_{i_2}^{-1} \le \theta$.

Definition 3.4 (*Support Extension,* [8, Sect. 2.A]) Let $S_p(\Xi)$ be a k-dimensional spline space on $[0, 1]$, and let Q be an element of the knot vector Ξ. We define the *support extension* $\tilde{Q}$ of Q by

$$\tilde{Q} := \left\{ \bigcup_{0 \le i < k} \operatorname{supp}(b_i^p) \colon b_i^p(x) \ne 0 \text{ for } x \in Q \right\}.$$

The same concept is generalised by tensor product construction to spline spaces on $[0, 1]^\ell$.

Assumption 3.5 (*Quasi-Uniformity of Knot Vectors*) All knot vectors will be assumed to be p-open and locally quasi-uniform, such that the usual spline theory of Beirão da Veiga et al. [7], Piegl and Tiller [3] and Schumaker [4] is applicable.

Throughout this paper, we will reserve the letter h for the maximal distance between two given knots and p for the minimal polynomial degree. Moreover $\tilde{h}$ will denote the maximal size of a support extension.

Definition 3.6 (*Patch*) We define a *patch* Γ to be the image of $\square$ under a diffeomorphism $\boldsymbol{F} \colon \square \to \Gamma \subseteq \mathbb{R}^3$. Let Ω be a Lipschitz domain. A *multipatch geometry* is defined to be a compact, orientable two-dimensional manifold $\Gamma = \partial\Omega$ invoked via $\Gamma = \bigcup_{0 \le j < N} \overline{\Gamma}_j$ by a family of patches $\{\Gamma_j\}_{0 \le j < N}$, $N \in \mathbb{N}$, given by a family of diffeomorphisms

$$\{\boldsymbol{F}_j \colon \square \hookrightarrow \Gamma_j\}_{0 \le j < N},$$

called *parametrisation*. We require the images of $\square$ of all $\boldsymbol{F}_j$ to be disjoint and that for any *patch interface* D of the form $D = \partial\Gamma_{j_0} \cap \partial\Gamma_{j_1} \ne \emptyset$, we find that the parametrisations $\boldsymbol{F}_{j_0}$ and $\boldsymbol{F}_{j_1}$ coincide up to orientation.

Note that this definition excludes non-watertight geometries and geometries with T-junctions, since mappings at interfaces must coincide.

In the spirit of isogeometric analysis, these mappings will usually be given by NURBS mappings, i.e., by

$$\boldsymbol{F}_j(x,y) := \sum_{0\le j_1<k_1}\sum_{0\le j_2<k_2} \frac{\boldsymbol{c}_{j_1,j_2} b_{j_1}^{p_1}(x) b_{j_2}^{p_2}(y) w_{j_1,j_2}}{\sum_{i_1=0}^{k_1-1}\sum_{i_2=0}^{k_2-1} b_{i_1}^{p_1}(x) b_{i_2}^{p_2}(y) w_{i_1,i_2}},$$

for control points $\boldsymbol{c}_{j_1,j_2} \in \mathbb{R}^3$ and weights $w_{i_1,i_2} > 0$. In accordance with the isogeometric framework, degrees and knot vectors of the discrete spaces to be mapped from the reference domain are usually chosen in accordance with the parametrisation [2]. However, the description of the geometry is independent of the analysis that will follow. From now on we reserve the letter N for the number of patches and the letter j to refer to a generic patch.

To avoid technical details, we introduce the following assumption.

Assumption 3.7 (*Smoothness of Geometry Mappings*) We assume any multipatch geometry to be given by an invertible, non-singular parametrisation $\{\boldsymbol{F}_j\}_{0\le j<N}$ that can be extended to mappings in $C^\infty(\overline{\square})$ such that the Jacobian of the extension has full rank everywhere on $\overline{\square}$.

Note that Assumption 3.7 implies that each patch Γ_j has a Lipschitz boundary. Moreover, we stress that, limited by the smoothness of $\{\boldsymbol{F}_j\}_{0\le j<N}$, variants of all results are still provable for non-smooth but invertible NURBS parametrisation. This would require an analysis via bent Sobolev spaces as in the works of Beirão da Veiga et al. [7]. Assumption 3.7 is merely for convenience. Moreover, it is possible to obtain parametric mappings satisfying Assumption 3.7 either through extraction of rational Bézier patches, which can be obtained as subpatches of a given NURBS parametrisation or, more generally, through an algorithmic approach as proposed by Harbrecht and Randrianarivony [9].

Definition 3.8 (*Spaces of Patchwise Regularity*) Let $\Gamma = \bigcup_{0\le j<N}\Gamma_j$ be a multipatch geometry. We define the norm

$$\|f\|_{H^s_{\mathrm{pw}}(\Gamma)}^2 := \sum_{0\le j<N} \|f|_{\Gamma_j}\|_{H^s(\Gamma_j)}^2$$

for $s > 0$ and all $f \in L^2(\Gamma)$ for which the right-hand side is well defined. We define the corresponding space equipped with this norm as

$$H^s_{\mathrm{pw}}(\Gamma) := \{f \in L^2(\Gamma) \colon \|f\|_{H^s_{\mathrm{pw}}(\Gamma)} < \infty\}.$$

In complete analogy, we extend the definition to vector-valued Sobolev spaces and spaces with graph norms. As usual, we denote these by bold letters $\boldsymbol{H}^s_{\mathrm{pw}}(\Gamma)$.

Definition 3.9 (*Single Patch Spline Complex,* [10]) Let $\boldsymbol{p} = (p_1, p_2)$ be a pair of positive integers and $\boldsymbol{\Xi} = (\Xi_1, \Xi_2)$ be a pair of p-open knot vectors on $[0, 1]$, where the multiplicity of interior knots in Ξ_j is at most p_j. Let Ξ_1' and Ξ_2' denote their truncation, i.e., the knot vector without its first and last knot. We define the *spline complex* on $\square$ as the spaces

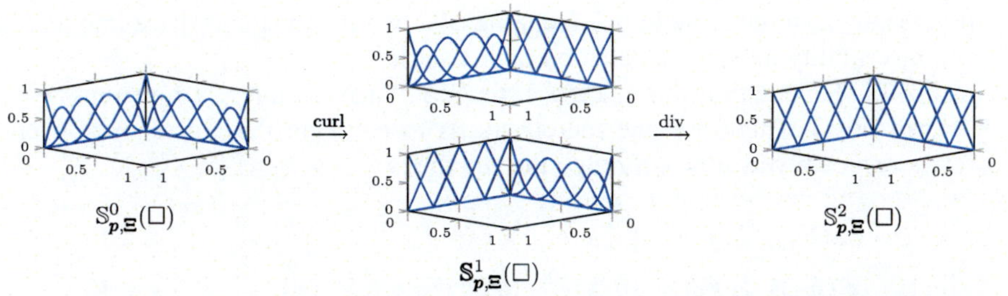

Fig. 3.2 Visualisation of the single patch spline-complex for $\boldsymbol{p} = (2, 2)$. The blue functions are the univariate B-splines related to each coordinate direction, whose tensor-product gives the bases of the spline spaces in Definition 3.9. The two-dimensional **curl** operator maps the smooth space to a vector valued space where the regularity in each vector component is lowered w.r.t. one spatial component. Analogously, the divergence operator maps to the space of globally lowered regularity

$$
\begin{aligned}
\mathbb{S}^0_{\boldsymbol{p},\boldsymbol{\Xi}}(\square) &:= S_{p_1,p_2}(\Xi_1, \Xi_2),\\
\boldsymbol{\mathbb{S}}^1_{\boldsymbol{p},\boldsymbol{\Xi}}(\square) &:= S_{p_1,p_2-1}(\Xi_1, \Xi_2') \times S_{p_1-1,p_2}(\Xi_1', \Xi_2),\\
\mathbb{S}^2_{\boldsymbol{p},\boldsymbol{\Xi}}(\square) &:= S_{p_1-1,p_2-1}(\Xi_1', \Xi_2').
\end{aligned}
$$

In the reference domain, the spline complex can be visualised as in Fig. 3.2. Assume Γ to consist of a single patch given via a geometry mapping $\boldsymbol{F}$ in accordance with Assumption 3.7. To define the spaces in the physical domain, we resort to an application of *pull-backs* and *push-forwards*.

For the case of two-dimensional mappings from $\square \to \Gamma_j$, Peterson [11] introduces the explicit form

$$
\begin{aligned}
\iota_0(\boldsymbol{F}_j)(f_0)(\boldsymbol{x}) &:= \big(f_0 \circ \boldsymbol{F}_j\big)(\boldsymbol{x}), && \boldsymbol{x} \in \square,\\
\iota_1(\boldsymbol{F}_j)(\boldsymbol{f}_1)(\boldsymbol{x}) &:= \big(\vartheta_{\boldsymbol{F}_j}(\boldsymbol{x}) \cdot (\boldsymbol{dF}_j)^{-1}(\boldsymbol{f}_1 \circ \boldsymbol{F}_j)\big)(\boldsymbol{x}), && \boldsymbol{x} \in \square,\\
\iota_2(\boldsymbol{F}_j)(f_2)(\boldsymbol{x}) &:= \big(\vartheta_{\boldsymbol{F}_j}(\boldsymbol{x}) \cdot (f_2 \circ \boldsymbol{F}_j)\big)(\boldsymbol{x}), && \boldsymbol{x} \in \square,
\end{aligned}
$$

where the term $\vartheta_{\boldsymbol{F}_j}$ for $\boldsymbol{x} \in \square$ is given by the so-called *surface measure*

$$
\vartheta_{\boldsymbol{F}_j}(\boldsymbol{x}) := \big\|\partial_{x_1}\boldsymbol{F}_j(\boldsymbol{x}) \times \partial_{x_2}\boldsymbol{F}_j(\boldsymbol{x})\big\|. \tag{3.2}
$$

The *integral-preserving* transformation ι_2 comes as a special case of the transformation used in (2.7). Note that if one were to compute the pullbacks $\iota_i(\boldsymbol{F}_j)$ for $i = 0, 1, 2$ as above, at first glance one were to encounter a dimensionality problem, since the inverse $\boldsymbol{dF}_j^{-1}$ of the Jacobian $\boldsymbol{dF}_j$ arising from $\boldsymbol{F}$ is of size 2×3, and thus not readily invertible. The work of e.g. Bossavit [12] and Kurz et al. [13] make it clear that, due to Assumption 3.7, required inverse mappings for the case of embedding a two-dimensional manifold into three-dimensional ambient space exist. They need to be understood as mappings from a tangent space of $\square$ onto a tangent space of Γ_j.

It is locally a smooth one to one mapping of a two-dimensional space into another, and invertibility must be understood in this sense.

For implementation this matters little. Both ansatz- and test functions will be defined on $\square$. Therefore one merely needs to compute the corresponding push-forwards, readily available through the equalities

$$\begin{aligned}
(\iota_0(\boldsymbol{F}_j))^{-1}(f_0) &= \big(f_0 \circ \boldsymbol{F}_j^{-1}\big)(\boldsymbol{x}), && \boldsymbol{x} \in \Gamma,\\
(\iota_1(\boldsymbol{F}_j))^{-1}(\boldsymbol{f}_1) &= \big(\vartheta(\boldsymbol{x})^{-1} \cdot (\boldsymbol{dF}_j)\boldsymbol{f}_1 \circ \boldsymbol{F}_j^{-1}\big)(\boldsymbol{x}), && \boldsymbol{x} \in \Gamma,\\
(\iota_2(\boldsymbol{F}_j))^{-1}(f_2) &= \big(\vartheta(\boldsymbol{x})^{-1} \cdot f_2 \circ \boldsymbol{F}_j^{-1}\big)(\boldsymbol{x}), && \boldsymbol{x} \in \Gamma.
\end{aligned}$$

That the corresponding inverse mappings are well-defined is guaranteed by Assumption 3.7. The inverse of $\boldsymbol{F}_j$ does not need to be computed, since $\boldsymbol{F}_j^{-1}(\Gamma) = \square$ is known by construction.

Remark 3.10 Peterson and Aberegg [14, Chap. 5] make clear that these mappings are conforming for $\Gamma_j \subseteq \mathbb{R}^3$, i.e., that the diagram

$$\begin{array}{ccccc}
H^1(\square) & \xrightarrow{\ \mathbf{curl}\ } & \boldsymbol{H}(\mathrm{div},\square) & \xrightarrow{\ \mathrm{div}\ } & L^2(\square)\\
\big\downarrow {\scriptstyle \iota_0(\boldsymbol{F}_j)^{-1}} & & \big\downarrow {\scriptstyle \iota_1(\boldsymbol{F}_j)^{-1}} & & \big\downarrow {\scriptstyle \iota_2(\boldsymbol{F}_j)^{-1}}\\
H^1(\Gamma_j) & \xrightarrow{\ \mathbf{curl}_\Gamma\ } & \boldsymbol{H}(\mathrm{div}_\Gamma,\Gamma_j) & \xrightarrow{\ \mathrm{div}_\Gamma\ } & L^2(\Gamma_j)
\end{array}$$

commutes. Because of this, we can identify the divergence on the reference domain with the divergence on the physical domain, up to a bounded factor induced by the corresponding pull-back. This is possible due to Assumption 3.7. We will, later on, utilise this explicitly to apply estimates of the kind

$$\|\mathrm{div}_\Gamma\, f\|_{L^2(\Gamma_j)} \simeq \big\|\mathrm{div}(f \circ \boldsymbol{F}_j)\big\|_{L^2(\square)},$$

see also the books of Monk [15] and Peterson [11] for a further review of these concepts.

Now we can define corresponding discretisations on the physical domain Γ_j by

$$\begin{aligned}
\mathbb{S}^0_{\boldsymbol{p},\boldsymbol{\Xi}}(\Gamma_j) &:= \big\{f : \iota_0(\boldsymbol{F}_j)(f) \in \mathbb{S}^0_{\boldsymbol{p},\boldsymbol{\Xi}}(\square)\big\},\\
\boldsymbol{\mathbb{S}}^1_{\boldsymbol{p},\boldsymbol{\Xi}}(\Gamma_j) &:= \big\{\boldsymbol{f} : \iota_1(\boldsymbol{F}_j)(\boldsymbol{f}) \in \boldsymbol{\mathbb{S}}^1_{\boldsymbol{p},\boldsymbol{\Xi}}(\square)\big\},\\
\mathbb{S}^2_{\boldsymbol{p},\boldsymbol{\Xi}}(\Gamma_j) &:= \big\{f : \iota_2(\boldsymbol{F}_j)(f) \in \mathbb{S}^2_{\boldsymbol{p},\boldsymbol{\Xi}}(\square)\big\}.
\end{aligned} \tag{3.3}$$

Proceeding as Beirão da Veiga et al. [7], the spline complex for spaces on the boundary is defined as follows.

Definition 3.11 (*Multipatch Spline Complex on Trace Spaces*) Let $\Gamma = \bigcup_{0\le j<N} \Gamma_j$ be a multipatch geometry satisfying Assumption 3.7. Moreover, let $\boldsymbol{\Xi} := (\boldsymbol{\Xi}_j)_{0\le j<N}$

be pairs of knot vectors in accordance with Assumption 3.5 and $\boldsymbol{p} = (\boldsymbol{p}_j)_{0 \leq j < N}$ pairs of integers, corresponding to polynomial degrees. Then we define the *spline complex on the boundary* Γ via

$$\mathbb{S}^0_{\boldsymbol{p},\boldsymbol{\Xi}}(\Gamma) := \left\{ f \in H^{1/2}(\Gamma) \colon f|_{\Gamma_j} \in \mathbb{S}^0_{\boldsymbol{p}_j,\boldsymbol{\Xi}_j}(\boldsymbol{\Gamma}_j) \text{ for all } 0 \leq j < N \right\},$$
$$\mathbf{S}^1_{\boldsymbol{p},\boldsymbol{\Xi}}(\Gamma) := \left\{ \boldsymbol{f} \in \boldsymbol{H}^{-1/2}_{\times}(\mathrm{div}_\Gamma, \Gamma) \colon \boldsymbol{f}|_{\Gamma_j} \in \mathbf{S}^1_{\boldsymbol{p}_j,\boldsymbol{\Xi}_j}(\boldsymbol{\Gamma}_j) \text{ for all } 0 \leq j < N \right\},$$
$$\mathbb{S}^2_{\boldsymbol{p},\boldsymbol{\Xi}}(\Gamma) := \left\{ f \in H^{-1/2}(\Gamma) \colon f|_{\Gamma_j} \in \mathbb{S}^2_{\boldsymbol{p}_j,\boldsymbol{\Xi}_j}(\boldsymbol{\Gamma}_j) \text{ for all } 0 \leq j < N \right\}.$$

We assume that $\boldsymbol{p}$ and $\boldsymbol{\Xi}$ coincide on every patch-interface.

Remark 3.12 Note that a different definition of the considered spline spaces can be given by application of the trace operators to the volumetric parametrisation, provided their existence, see Theorem 2.26. However, the construction above seems more suitable for the analysis of approximation properties. Moreover, the above easily generalises to screens and non-orientable surfaces.

3.2 Approximation Properties of Conforming Spline Spaces

We will now investigate approximation properties of the spaces defined in the previous section. This will be done through the introduction of quasi-interpolation operators, projections, which are defined in terms of a dual basis.

For one-dimensional spline spaces Schumaker [4, Sect. 4.2] introduced quasi-interpolants

$$\Pi_{p,\Xi} \colon f \mapsto \sum_{0 \leq i < k} \lambda_{i,p}(f) b_i^p.$$

They are defined in terms of dual functionals

$$\lambda_{i,p} \colon L^2(\xi_i, \xi_{i+p+1}) \to \mathbb{K},$$
$$\lambda_{i,p} \colon f \mapsto \int_{\xi_i}^{\xi_{i+p+1}} f(x) \partial^{p+1} \psi_i(x) \,\mathrm{d}x,$$

for a weight function ψ_i only depending on p and the knot vector Ξ. By definition of the $\lambda_{i,p}$, they merely require f to be square integrable.

From these quasi interpolation operators one can derive projection operators $\Pi^0_{\boldsymbol{p},\boldsymbol{\Xi}}$, $\boldsymbol{\Pi}^1_{\boldsymbol{p},\boldsymbol{\Xi}}$, $\Pi^2_{\boldsymbol{p},\boldsymbol{\Xi}}$ mapping onto the spaces $\mathbb{S}^0_{\boldsymbol{p},\boldsymbol{\Xi}}(\square)$, $\mathbf{S}^1_{\boldsymbol{p},\boldsymbol{\Xi}}(\square)$, and $\mathbb{S}^2_{\boldsymbol{p},\boldsymbol{\Xi}}(\square)$, see [7, p. 169ff], where error estimations and L^2-stability for B-spline approximations have been also been provided. A crucial property of these operators is the following.

Lemma 3.13 (Commuting Interpolation Operators, [7, Proposition 5.8]) *The diagram*

$$\begin{array}{ccccc} H^1(\square) & \xrightarrow{\textbf{curl}} & \boldsymbol{H}(\mathrm{div},\square) & \xrightarrow{\mathrm{div}} & L^2(\square) \\ \downarrow \Pi^0_{\boldsymbol{p},\boldsymbol{\Xi}} & & \downarrow \boldsymbol{\Pi}^1_{\boldsymbol{p},\boldsymbol{\Xi}} & & \downarrow \Pi^2_{\boldsymbol{p},\boldsymbol{\Xi}} \\ \mathbb{S}^0_{\boldsymbol{p},\boldsymbol{\Xi}}(\square) & \xrightarrow{\textbf{curl}} & \boldsymbol{\mathbb{S}}^1_{\boldsymbol{p},\boldsymbol{\Xi}}(\square) & \xrightarrow{\mathrm{div}} & \mathbb{S}^2_{\boldsymbol{p},\boldsymbol{\Xi}}(\square) \end{array}$$

commutes.

Remark 3.14 For the two-dimensional setting Beirão da Veiga et al. [7, Sect. 5.5] define two spaces $\boldsymbol{\mathbb{S}}^1_{\boldsymbol{p},\boldsymbol{\Xi}}$ and $\boldsymbol{\mathbb{S}}^{1*}_{\boldsymbol{p},\boldsymbol{\Xi}}$, which correspond to curl conforming and divergence conforming spaces, respectively. Since we are interested mostly in spaces of the div-type and the spaces differ only by a rotation, we will not mention the two different types of spline spaces. However, it should be noted that our spaces of type $\boldsymbol{\mathbb{S}}^1_{\boldsymbol{p},\boldsymbol{\Xi}}$ correspond to those of type $\boldsymbol{\mathbb{S}}^{1*}_{\boldsymbol{p},\boldsymbol{\Xi}}$ in the cited literature.

By application of the pull-backs used to define the spline spaces one can immediately generalise the projectors and all results to the case of functions on the physical domains. Beirão da Veiga et al. [7, Cor. 5.12] show that for the case of a single patch Γ_j the following holds.

Corollary 3.15 (Single Patch Approximation Estimate, [7, Cor. 5.12]) *Let Γ_j be a single patch domain and let Assumptions 3.5 and 3.7 hold. Then we find that*

$$\begin{aligned} \left\| u - \Pi^0_{\boldsymbol{p},\boldsymbol{\Xi}} u \right\|_{H^r(\Gamma_j)} &\lesssim h^{s-r} \|u\|_{H^s(\Gamma_j)}, \quad \text{for } 0 \le r \le s \le p+1, \\ \left\| \boldsymbol{u} - \boldsymbol{\Pi}^1_{\boldsymbol{p},\boldsymbol{\Xi}} \boldsymbol{u} \right\|_{\boldsymbol{H}^r(\Gamma_j)} &\lesssim h^{s-r} \|\boldsymbol{u}\|_{\boldsymbol{H}^s(\Gamma_j)}, \quad \text{for } 0 \le r \le s \le p, \\ \left\| u - \Pi^2_{\boldsymbol{p},\boldsymbol{\Xi}} u \right\|_{H^r(\Gamma_j)} &\lesssim h^{s-r} \|u\|_{H^s(\Gamma_j)}, \quad \text{for } 0 \le r \le s \le p. \end{aligned}$$

By properties of the pull-backs, the operators also commute w.r.t. the surface differential operators. Thus, one finds that

$$\begin{aligned} \left\| \boldsymbol{u} - \boldsymbol{\Pi}^1_{\boldsymbol{p},\boldsymbol{\Xi}} \boldsymbol{u} \right\|^2_{\boldsymbol{H}^r(\mathrm{div}_\Gamma,\Gamma_j)} &= \left\| \boldsymbol{u} - \boldsymbol{\Pi}^1_{\boldsymbol{p},\boldsymbol{\Xi}} \boldsymbol{u} \right\|^2_{\boldsymbol{H}^r(\Gamma_j)} + \left\| \mathrm{div}_\Gamma(\boldsymbol{u}) - \mathrm{div}_\Gamma(\boldsymbol{\Pi}^1_{\boldsymbol{p},\boldsymbol{\Xi}} \boldsymbol{u}) \right\|^2_{H^r(\Gamma_j)} \\ &= \left\| \boldsymbol{u} - \boldsymbol{\Pi}^1_{\boldsymbol{p},\boldsymbol{\Xi}} \boldsymbol{u} \right\|^2_{\boldsymbol{H}^r(\Gamma_j)} + \left\| \mathrm{div}_\Gamma(\boldsymbol{u}) - \Pi^2_{\boldsymbol{p},\boldsymbol{\Xi}} \mathrm{div}_\Gamma(\boldsymbol{u}) \right\|^2_{H^r(\Gamma_j)}. \end{aligned}$$

This allows to apply the estimates of the previous corollary, which yields

$$\left\| \boldsymbol{u} - \boldsymbol{\Pi}^1_{\boldsymbol{p},\boldsymbol{\Xi}} \boldsymbol{u} \right\|_{\boldsymbol{H}^r(\mathrm{div}_\Gamma,\Gamma_j)} \lesssim h^{s-r} \|\boldsymbol{u}\|_{\boldsymbol{H}^s(\mathrm{div}_\Gamma,\Gamma_j)}, \qquad \text{for } 0 \le r \le s \le p.$$

In the following, we will generalise these notions for the multipatch case.

3.2.1 Multipatch Quasi-Interpolation Operators

To obtain a generalisation of the previous result to the multipatch setting, we first construct interpolation operators that can preserve continuity across patch boundaries. For one-dimensional spline spaces $S_p(\Xi)$ and $f \in C^\infty(0, 1)$, Beirão da Veiga et al. [7, Sect. 2.1.5] define

$$\tilde{\Pi}_{p,\Xi} \colon f \mapsto \sum_{0 \le i < k} \tilde{\lambda}_{i,p}(f) b_i^p,$$

where for $0 < i < k-1$ we set $\tilde{\lambda}_{i,p}(f) = \lambda_{i,p}(f)$, but additionally, require

$$\tilde{\lambda}_{0,p}(f) = f(0) \quad \text{as well as} \quad \tilde{\lambda}_{k-1,p}(f) = f(1). \tag{3.4}$$

This will yield versions of the projection operators which respect boundary conditions.

Note that $\tilde{\lambda}_{0,p}$ and $\tilde{\lambda}_{k-1,p}$ are well defined for continuous input, for which point evaluations are definable. Noting this, for some suitable domain $\mathrm{Dom}(\tilde{\Pi}_{p,\Xi})$ to be investigated later on we can construct quasi-interpolation operators for the multipatch case that commute w.r.t. derivation, analogously to the construction of Buffa et al. [10]. Investigation of the one-dimensional diagram

$$\begin{array}{ccc} \mathrm{Dom}(\tilde{\Pi}_{p,\Xi}) & \xrightarrow{\partial_x} & \mathrm{Dom}(\tilde{\Pi}^{\partial}_{p,\Xi}) \\ \downarrow \tilde{\Pi}_{p,\Xi} & & \downarrow \tilde{\Pi}^{\partial}_{p,\Xi} \\ S_p(\Xi) & \xrightarrow{\partial_x} & S_{p-1}(\Xi') \end{array} \tag{3.5}$$

makes clear that a suitable choice of $\tilde{\Pi}^{\partial}_{p,\Xi}$ is given by

$$\tilde{\Pi}^{\partial}_{p,\Xi} \colon f \mapsto \partial_\eta \left[\tilde{\Pi}_{p,\Xi} \int_0^\eta f(x)\, \mathrm{d}x \right]. \tag{3.6}$$

By diagram chase and application of the fundamental theorem of calculus one can see that (3.6) renders Diagram (3.5) commutative.

While $\tilde{\Pi}_{p,\Xi}$ has been investigated already [8], $\tilde{\Pi}^{\partial}_{p,\Xi}$ and all consequential concepts have not been analysed. We will start with the following essential observation.

Proposition 3.16 (Spline Preserving Property) *The operator*

$$\tilde{\Pi}^{\partial}_{p,\Xi} \colon L^2(0,1) \to S^{p-1}(\Xi')$$

preserves B-splines within $S^{p-1}(\Xi')$.

Proof From the results of Buffa et al. [8, Sect. 2] we know that the assertion holds for $\tilde{\Pi}_{p,\Xi}$. Fixing a spline $b' \in S^{p-1}(\Xi')$, we know that there exists a spline $b \in S^p(\Xi)$ with $\partial_x b = b'$, since $\partial_x \colon S^p(\Xi) \to S^{p-1}(\Xi')$ is surjective. Now, since b is within the domain of the interpolation operator, the assertion follows by diagram chase. ■

An immediate consequence of this proposition is that the operator $\tilde{\Pi}^{\partial}_{p,\Xi}$ is a projection. Moreover, it can be safely applied to functions in $H^1(0,1)$, as shown by the following proposition.

Proposition 3.17 (Stability of $\tilde{\Pi}_{p,\Xi}$) *Let Assumption 3.5 hold. Assume f to be continuous in a neighbourhood around 0 and 1 and let $I = (\xi_j, \xi_{j+1})$. Let $\tilde{I}$ denote the support extension of I. Then it holds that*

$$\left\|\tilde{\Pi}_{p,\Xi}(f)\right\|_{L^2(I)} \lesssim \|f\|_{L^2(\tilde{I})} + h|f|_{H^1(\tilde{I})}, \tag{3.7}$$

$$\left|\tilde{\Pi}_{p,\Xi}(f)\right|_{H^1(I)} \lesssim \|f\|_{H^1(\tilde{I})}. \tag{3.8}$$

Moreover, we find

$$\|\tilde{\Pi}^{\partial}_{p,\Xi}(f)\|_{L^2(I)} \lesssim \|f\|_{L^2(\tilde{I})}.$$

Proof The first two inequalities have been discussed by Buffa et al. [8, Proposition 2.3]. Investigating the third assertion, we set $g(x) = \int_0^x f(t)\,\mathrm{d}t$.

The proof concludes by a nontrivial application of the Poincaré inequality. Set $C = -\frac{1}{|\tilde{I}|}\int_{\tilde{I}} g\,\mathrm{d}x$, where $\left|\tilde{I}\right|$ denotes the Lebesgue measure of $\tilde{I}$, and observe that

$$\begin{aligned}
\|\tilde{\Pi}^{\partial}_{p,\Xi}(f)\|_{L^2(I)} &= \left\|\partial_x\tilde{\Pi}_{p,\Xi}\int_0^x f(t)\,\mathrm{d}t\right\|_{L^2(I)}\\
&= \left\|\partial_x\tilde{\Pi}_{p,\Xi}\left(\int_0^x f(t)\,\mathrm{d}t + C\right)\right\|_{L^2(I)}\\
&= \left|\tilde{\Pi}_{p,\Xi}\left(\int_0^x f(t)\,\mathrm{d}t + C\right)\right|_{H^1(I)}\\
&\lesssim \left(\|g+C\|^2_{L^2(\tilde{I})} + |g+C|^2_{H^1(\tilde{I})}\right)^{1/2},
\end{aligned}$$

where the inequality follows from (3.8). We will now estimate the two terms in the last line separately.

By definition of C we find that $\frac{1}{|\tilde{I}|}\int_{\tilde{I}} g\,\mathrm{d}x = -C$ and can apply Lemma 2.14. This yields

$$\|g+C\|_{L^2(\tilde{I})} \lesssim |g|_{H^1(\tilde{I})} = \|f\|_{L^2(\tilde{I})}.$$

For the second term we find

$$|g + C|^2_{H^1(\tilde{I})} = \int_{\tilde{I}} |\partial_x(g(x) + C)|^2 \,\mathrm{d}x = |g|^2_{H^1(\tilde{I})} = \|f\|^2_{L^2(\tilde{I})}.$$

Combining all estimates the assertion follows. It should be remarked that the constants in the estimates involving Lemma 2.14 can be chosen independently of h, since bounds for the Poincaré constant on the unit interval are known [16] and a restriction to subintervals cannot increase it [17, Lemma 4]. ■

For the exact domains of definitions $\mathrm{Dom}(\tilde{\Pi}_{p,\Xi})$ and $\mathrm{Dom}(\tilde{\Pi}^{\partial}_{p,\Xi})$ of the projection operators, the last proposition implies that

$$\begin{aligned} \mathrm{Dom}(\tilde{\Pi}_{p,\Xi}) &\supseteq H^1(0,1), \\ \mathrm{Dom}(\tilde{\Pi}^{\partial}_{p,\Xi}) &\supseteq L^2(0,1). \end{aligned}$$

We define quasi-interpolation operators via

$$\begin{aligned} \tilde{\Pi}^0_{\boldsymbol{p},\boldsymbol{\Xi}} &:= \tilde{\Pi}_{p_1,\Xi_1} \otimes \tilde{\Pi}_{p_2,\Xi_2}, \\ \tilde{\boldsymbol{\Pi}}^1_{\boldsymbol{p},\boldsymbol{\Xi}} &:= (\tilde{\Pi}_{p_1,\Xi_1} \otimes \tilde{\Pi}^{\partial}_{p_2,\Xi_2}) \times (\tilde{\Pi}^{\partial}_{p_1,\Xi_1} \otimes \tilde{\Pi}_{p_2,\Xi_2}), \\ \tilde{\Pi}^2_{\boldsymbol{p},\boldsymbol{\Xi}} &:= \tilde{\Pi}^{\partial}_{p_1,\Xi_1} \otimes \tilde{\Pi}^{\partial}_{p_2,\Xi_2}. \end{aligned} \tag{3.9}$$

Proposition 3.17 together with the tensor product construction (3.9) makes it clear that suitable domains of the interpolation operators are given by the tensor product spaces

$$\mathrm{Dom}(\tilde{\Pi}^0_{\boldsymbol{p},\boldsymbol{\Xi}}) \supseteq H^1(0,1) \otimes H^1(0,1), \tag{3.10}$$

$$\mathrm{Dom}(\tilde{\boldsymbol{\Pi}}^1_{\boldsymbol{p},\boldsymbol{\Xi}}) \supseteq \big(H^1(0,1) \otimes L^2(0,1)\big) \times \big(L^2(0,1) \otimes H^1(0,1)\big), \tag{3.11}$$

$$\mathrm{Dom}(\tilde{\Pi}^2_{\boldsymbol{p},\boldsymbol{\Xi}}) \supseteq L^2(0,1) \otimes L^2(0,1) = L^2(\square). \tag{3.12}$$

In this sense, we have the following.

Lemma 3.18 (Commutativity in the Reference Domain) *The diagram*

$$\begin{array}{ccccc} \mathrm{Dom}(\tilde{\Pi}^0_{\boldsymbol{p},\boldsymbol{\Xi}}) & \xrightarrow{\ \mathbf{curl}\ } & \mathrm{Dom}(\tilde{\boldsymbol{\Pi}}^1_{\boldsymbol{p},\boldsymbol{\Xi}}) & \xrightarrow{\ \mathrm{div}\ } & \mathrm{Dom}(\tilde{\Pi}^2_{\boldsymbol{p},\boldsymbol{\Xi}}) \\ \big\downarrow \tilde{\Pi}^0_{\boldsymbol{p},\boldsymbol{\Xi}} & & \big\downarrow \tilde{\boldsymbol{\Pi}}^1_{\boldsymbol{p},\boldsymbol{\Xi}} & & \big\downarrow \tilde{\Pi}^2_{\boldsymbol{p},\boldsymbol{\Xi}} \\ \mathbb{S}^0_{\boldsymbol{p},\boldsymbol{\Xi}}(\square) & \xrightarrow{\ \mathbf{curl}\ } & \boldsymbol{\mathbb{S}}^1_{\boldsymbol{p},\boldsymbol{\Xi}}(\square) & \xrightarrow{\ \mathrm{div}\ } & \mathbb{S}^2_{\boldsymbol{p},\boldsymbol{\Xi}}(\square) \end{array}$$

commutes.

Proof By (3.5) we have $\tilde{\Pi}^{\partial}_{p,\Xi} \circ \partial_x = \partial_x \circ \tilde{\Pi}_{p,\Xi}$. The characterisation of the domains and the definition of the interpolation operators via tensor products now yields the assertion by application of the differential operators in their representation w.r.t. cartesian coordinates and simple computations, in analogy to [7, Proposition 5.8]. ■

We can now define global projections on the physical domain via application of the pull-backs.

Definition 3.19 (*Global Interpolation Operators*) Let $\boldsymbol{\Xi}$ and $\boldsymbol{p}$ denote N-tuples of pairs of knot vectors and polynomial degrees, respectively. Let $\Gamma = \bigcup_{0\leq j<N}\Gamma_j$ be a geometry induced by a family of diffeomorphisms $\{\boldsymbol{F}_j\}_{0\leq j<N}$ as in Definition 3.6. For a family of patchwise linear operators $\{L_j\}_{0\leq j<N}$ we denote by $\{\tilde{L}_j\}_{0\leq j<N}$ their extensions by 0 onto Γ and write

$$\bigoplus_{0\leq j<N} L_j := \sum_{0\leq j<N} \tilde{L}_j.$$

Now we define the global B-spline projections as

$$\begin{aligned}
\tilde{\Pi}^0_\Gamma &:= \bigoplus_{0\leq j<N}\Big((\iota_0(\boldsymbol{F}_j))^{-1}\circ \tilde{\Pi}^0_{\boldsymbol{p}_j,\boldsymbol{\Xi}_j}\circ \iota_0(\boldsymbol{F}_j)\Big),\\
\tilde{\boldsymbol{\Pi}}^1_\Gamma &:= \bigoplus_{0\leq j<N}\Big((\iota_1(\boldsymbol{F}_j))^{-1}\circ \tilde{\boldsymbol{\Pi}}^1_{\boldsymbol{p}_j,\boldsymbol{\Xi}_j}\circ \iota_1(\boldsymbol{F}_j)\Big),\\
\tilde{\Pi}^2_\Gamma &:= \bigoplus_{0\leq j<N}\Big((\iota_2(\boldsymbol{F}_j))^{-1}\circ \tilde{\Pi}^2_{\boldsymbol{p}_j,\boldsymbol{\Xi}_j}\circ \iota_2(\boldsymbol{F}_j)\Big),
\end{aligned}$$

i.e., by patchwise application of the projections of (3.9) with their corresponding pull-backs and push-forwards.

Note that, since the pull-backs are commuting with the differential operators in the reference domain and surface differential operators [11], an analogue of Lemma 3.13 holds also for the global operators.

Moreover, an analogue characterisation of the domain of these global operators to that of Eqs. (3.10)–(3.12) is possible. However, since this will be technically involved, we will rely on relation (2.5) and conduct the upcoming approximation analysis with respect to the domains $\mathrm{Dom}(\tilde{\Pi}^0_\Gamma) = H^2_{\mathrm{pw}}(\Gamma)$, $\mathrm{Dom}(\tilde{\boldsymbol{\Pi}}^1_\Gamma) = \boldsymbol{H}^1_{\mathrm{pw}}(\Gamma)$, and $\mathrm{Dom}(\tilde{\Pi}^2_\Gamma) = L^2(\Gamma)$, respectively. This yields the following.

Lemma 3.20 (Commutativity in the Physical Domain) *The diagram*

$$\begin{array}{ccccc}
\mathrm{Dom}(\tilde{\Pi}^0_\Gamma) & \xrightarrow{\mathrm{curl}_\Gamma} & \mathrm{Dom}(\tilde{\boldsymbol{\Pi}}^1_\Gamma) & \xrightarrow{\mathrm{div}_\Gamma} & \mathrm{Dom}(\tilde{\Pi}^2_\Gamma)\\
\downarrow{\scriptstyle \tilde{\Pi}^0_\Gamma} & & \downarrow{\scriptstyle \tilde{\boldsymbol{\Pi}}^1_\Gamma} & & \downarrow{\scriptstyle \tilde{\Pi}^2_{\boldsymbol{p},\boldsymbol{\Xi}}}\\
\mathbb{S}^0_{\boldsymbol{p},\boldsymbol{\Xi}}(\Gamma) & \xrightarrow{\mathrm{curl}_\Gamma} & \boldsymbol{\mathbb{S}}^1_{\boldsymbol{p},\boldsymbol{\Xi}}(\Gamma) & \xrightarrow{\mathrm{div}_\Gamma} & \mathbb{S}^2_{\boldsymbol{p},\boldsymbol{\Xi}}(\Gamma)
\end{array}$$

commutes.

The construction of (3.9) and Definition 3.19 can easily be generalised to three dimensions, see Appendix A.

3.2.2 Approximation Estimates

We will now provide approximation estimates for the introduced interpolation operators. Note that, by construction, it is clear that the boundary interpolating projections commute w.r.t. the differential operators. It is however not clear whether the condition (3.4) and the construction in (3.6) and (3.9) impacts the convergence behaviour w.r.t. h-refinement.

To utilise the commuting property to show convergence in the energy spaces, we need an analogue of Corollary 3.15 for the multipatch operators.

Utilising the stability assertion in Proposition 3.17, we now can provide an error estimate in one dimension.

Proposition 3.21 (Approximation Properties of $\tilde{\Pi}_{p,\Xi}$) *Let the assumptions of Proposition 3.17 hold. For integers $1 \leq s \leq p+1$ one finds*

$$\left\| f - \tilde{\Pi}_{p,\Xi} f \right\|_{L^2(I)} \lesssim h^s \|f\|_{H^s(\tilde{I})}, \qquad \text{for all } f \in H^s(0,1),$$

and for integers $0 \leq s \leq p$ one finds

$$\| f - \tilde{\Pi}^{\partial}_{p,\Xi} f \|_{L^2(I)} \lesssim h^s \|f\|_{H^s(\tilde{I})}, \qquad \text{for all } f \in H^s(0,1).$$

Proof We investigate merely the case of $\tilde{\Pi}_{p,\Xi}$. Due to the stability of $\tilde{\Pi}^{\partial}_{p,\Xi}$ as discussed in Proposition 3.17, we can prove the other case by similar means.

For the first inequality, as in [7, Proposition 4.2], it is enough to consider classical polynomial estimates together with Proposition 3.17 to achieve

$$\begin{aligned}
\left\| f - \tilde{\Pi}_{p,\Xi} f \right\|_{L^2(I)} &\leq \|f - q\|_{L^2(I)} + \left\| \tilde{\Pi}_{p,\Xi}(q - f) \right\|_{L^2(I)} \\
&\lesssim \|f - q\|_{L^2(I)} + \|q - f\|_{L^2(\tilde{I})} + h|q - f|_{H^1(\tilde{I})} \\
&\lesssim h^s \|f\|_{H^s(\tilde{I})},
\end{aligned}$$

for the best approximating polynomial q of degree no higher than p. Herein, the last inequality holds by classical polynomial estimates, see e.g. [18, Lemma 4.3.8]. ■

We can now state the main result of this section. Its proof is a standard tensor product argument, cf. Sect. 2.4 and the literature cited therein. While by Definition 2.22 tensor product operators can be applied to arbitrary elements of the tensor product space, we will conduct the upcoming proof on the elementary level of finite linear combinations of simple tensors, cf. Definition 2.19. This is done for the readers convenience, since these foundational arguments are usually not made transparent in the literature concerned with isogeometric analysis.

Theorem 3.22 (Approximation via Commuting Multipatch Quasi-Interpolants) *Let Assumptions 3.5 and 3.7 be satisfied and let s be integer-valued. Let $f_0 \in H^s_{\mathrm{pw}}(\Gamma)$, 2*

$\leq s$, as well as $\boldsymbol{f}_1 \in \boldsymbol{H}^s_{\mathrm{pw}}(\Gamma)$, $1 \leq s$, and $f_2 \in H^s_{\mathrm{pw}}(\Gamma)$, $0 \leq s$. Moreover, let each function be within the domain of the interpolation operator applied below, cf. Lemma 3.20. We find that

$$\begin{aligned}
\left\| f_0 - \tilde{\Pi}^0_\Gamma f_0 \right\|_{L^2(\Gamma)} &\lesssim h^s \| f_0 \|_{H^s_{\mathrm{pw}}(\Gamma)}, && \text{for } 2 \leq s \leq p+1,\\
\left\| f_0 - \tilde{\Pi}^0_\Gamma f_0 \right\|_{H^1(\Gamma)} &\lesssim h^{s-1} \| f_0 \|_{H^s_{\mathrm{pw}}(\Gamma)}, && \text{for } 2 \leq s \leq p+1,\\
\| \boldsymbol{f}_1 - \tilde{\boldsymbol{\Pi}}^1_\Gamma \boldsymbol{f}_1 \|_{\boldsymbol{L}^2(\Gamma)} &\lesssim h^s \| \boldsymbol{f}_1 \|_{\boldsymbol{H}^s_{\mathrm{pw}}(\Gamma)}, && \text{for } 1 \leq s \leq p,\\
\left\| f_2 - \tilde{\Pi}^2_\Gamma f_2 \right\|_{L^2(\Gamma)} &\lesssim h^s \| f_2 \|_{H^s_{\mathrm{pw}}(\Gamma)}, && \text{for } 0 \leq s \leq p.
\end{aligned}$$

Furthermore, the operators satisfy the estimate

$$\| \boldsymbol{f}_1 - \tilde{\boldsymbol{\Pi}}^1_\Gamma \boldsymbol{f}_1 \|_{\boldsymbol{H}^0(\mathrm{div}_\Gamma, \Gamma)} \lesssim h^s \| \boldsymbol{f}_1 \|_{\boldsymbol{H}^s_{\mathrm{pw}}(\mathrm{div}_\Gamma, \Gamma)}, \qquad \text{for } 1 \leq s \leq p. \tag{3.13}$$

Proof Due to Assumption 3.7 and the locality of all involved norms we can pull all quantities back to $\square$. Thus, it suffices to provide a patchwise argument in the reference domain. Note that the regularity of the spline approximation is always sufficient for the involved norms to be defined since it is enforced by the interpolation property of the $\tilde{\Pi}$ at the patch interfaces.

Buffa et al. [8, Proposition 4.2] directly provide

$$\left\| f - \tilde{\Pi}_{p,\Xi} f \right\|_{H^r(\square)} \lesssim h^{s-r} \| f \|_{H^s(\square)}, \tag{3.14}$$

for $r = 0, 1$, from which the $\tilde{\Pi}^0_\Gamma$ case follows immediately.

We will now provide a proof for the $\tilde{\Pi}^2_\Gamma$ case by investigating $\tilde{\Pi}^2_{\boldsymbol{p},\boldsymbol{\Xi}} = \tilde{\Pi}^\partial_{p,\Xi_1} \otimes \tilde{\Pi}^\partial_{p,\Xi_2}$, which will be done largely analogous to the proofs within the cited literature.

The third assertion, i.e., the case of $\tilde{\boldsymbol{\Pi}}^1_\Gamma$ follows from a combination of the arguments discussed here and in [8, Proposition 4.2].

In the upcoming proof, we will make the tensor product structure as discussed in Sect. 2.4 explicit. Let, for now, $f = \sum_{i,i'<M} c_{i,i'} f_i f_{i'}$ be a finite linear combination of simple tensors from $H^s(0,1) \otimes H^s(0,1) \supseteq H^{2s}(\square)$ for integers $s \geq 0$. Note that by Definition 2.19 and linearity of $\tilde{\Pi}^2_\Gamma$ it suffices to show the property for such f and then apply suitable density arguments. Let $I_1 \times I_2 = Q \subset \square$ be an element. One can estimate via triangle inequality that

$$\begin{aligned}
\| f - \tilde{\Pi}^2_{\boldsymbol{p},\boldsymbol{\Xi}} f \|_{L^2(Q)} &= \| f - (\tilde{\Pi}^\partial_{p,\Xi_1} \otimes \tilde{\Pi}^\partial_{p,\Xi_2})(f) \|_{L^2(Q)}\\
&\leq \| f - (\tilde{\Pi}^\partial_{p,\Xi_1} \otimes \mathrm{Id})(f) \|_{L^2(Q)}\\
&\quad + \| (\tilde{\Pi}^\partial_{p,\Xi_1} \otimes \mathrm{Id})(f) - (\tilde{\Pi}^\partial_{p,\Xi_1} \otimes \tilde{\Pi}^\partial_{p,\Xi_2})(f) \|_{L^2(Q)}.
\end{aligned} \tag{3.15}$$

A simple Gram–Schmidt argument makes clear that, since f is a finite linear combination of simple tensors, we may assume without loss of generality that f is represented by

$$f = \sum_{i,i'<M} c_{i,i'} \varphi_i \otimes \psi_{i'},$$

where $\{\varphi_i|_{I_1}\}$ are $L^2(I_1)$-orthogonal and the $\{\psi_i|_{I_2}\}$ are $L^2(I_2)$-orthogonal.

By Proposition 3.21 we can immediately estimate the first term of (3.15) via

$$\begin{aligned}
&\left\| f - (\tilde{\Pi}^{\partial}_{p,\Xi_1} \otimes \mathrm{Id}) f \right\|^2_{L^2(Q)} && (3.16)\\
&= \langle f - (\tilde{\Pi}^{\partial}_{p,\Xi_1} \otimes \mathrm{Id}) f, f - (\tilde{\Pi}^{\partial}_{p,\Xi_1} \otimes \mathrm{Id}) f \rangle_{L^2(Q)}\\
&= \left\langle \sum_{i,i'<M} c_{i,i'} (\varphi_i - \tilde{\Pi}^{\partial}_{p,\Xi_1} \varphi_i) \otimes \psi_{i'}, \sum_{j,j'<M} c_{j,j'} (\varphi_j - \tilde{\Pi}^{\partial}_{p,\Xi_1} \varphi_j) \otimes \psi_{j'} \right\rangle_{L^2(Q)}\\
&= \sum_{i,i'<M} c_{i,i'} \sum_{j,j'<M} c_{j,j'} \langle (\varphi_i - \tilde{\Pi}^{\partial}_{p,\Xi_1} \varphi_i) \otimes \psi_{i'}, (\varphi_j - \tilde{\Pi}^{\partial}_{p,\Xi_1} \varphi_j) \otimes \psi_{j'} \rangle_{L^2(Q)}\\
&= \sum_{i,i'<M} c_{i,i'} \sum_{j,j'<M} c_{j,j'} \langle (\varphi_i - \tilde{\Pi}^{\partial}_{p,\Xi_1} \varphi_i), (\varphi_j - \tilde{\Pi}^{\partial}_{p,\Xi_1} \varphi_j) \rangle_{L^2(I_1)} \langle \psi_{i'}, \psi_{j'} \rangle_{L^2(I_2)}\\
&= \sum_{i,i'<M} c_{i,i'} \langle (\varphi_i - \tilde{\Pi}^{\partial}_{p,\Xi_1} \varphi_i), (\varphi_i - \tilde{\Pi}^{\partial}_{p,\Xi_1} \varphi_i) \rangle_{L^2(I_1)} \langle \psi_{i'}, \psi_{i'} \rangle_{L^2(I_2)} && (3.17)\\
&\lesssim \sum_{i,i'<M} c_{i,i'} h^{2s} \langle \varphi_i, \varphi_i \rangle_{H^s(\tilde{I}_1)} \langle \psi_{i'}, \psi_{i'} \rangle_{L^2(I_2)}\\
&= \sum_{i,i'<M} c_{i,i'} \sum_{j,j'<M} c_{j,j'} h^{2s} \langle \varphi_i, \varphi_j \rangle_{H^s(\tilde{I}_1)} \langle \psi_{i'}, \psi_{j'} \rangle_{L^2(I_2)} && (3.18)\\
&= \sum_{i,i'<M} c_{i,i'} \sum_{j,j'<M} c_{j,j'} h^{2s} \langle \varphi_i \otimes \psi_{i'}, \varphi_j \otimes \psi_{j'} \rangle_{H^s(\tilde{I}_1) \otimes L^2(I_2)}\\
&= h^{2s} \langle f, f \rangle_{H^s(\tilde{I}_1) \otimes L^2(I_2)}\\
&= h^{2s} \| f \|^2_{H^s(\tilde{I}_1) \otimes L^2(I_2)}. && (3.19)
\end{aligned}$$

Therein, the orthogonality is used in Eqs. (3.17) and (3.18) where all cross-terms vanish. We can estimate the second term of (3.15) by similar means, utilising the stability property from Propositions 3.17 and 3.21, which yields

$$\| (\tilde{\Pi}^{\partial}_{p,\Xi_1} \otimes \mathrm{Id})(f) - (\tilde{\Pi}^{\partial}_{p,\Xi_1} \otimes \tilde{\Pi}^{\partial}_{p,\Xi_2})(f) \|^2_{L^2(Q)} \lesssim h^{2s} \| f \|^2_{H^s(\tilde{I}_1) \otimes L^2(I_2)}. \tag{3.20}$$

Combining the above with (3.15) yields

$$\| f - \tilde{\Pi}^2_{\boldsymbol{p},\boldsymbol{\Xi}} f \|^2_{L^2(Q)} \lesssim h^{2s} \| f \|^2_{H^s(\tilde{Q})}. \tag{3.21}$$

The result transfers to $H^s(0,1) \otimes H^s(0,1) \supseteq H^{2s}(\square)$ by density of finite linear combinations of simple tensors, cf. Definition 2.19. Now density of $H^{2s}(\square)$ in $H^s(\square)$ yields the assertion, since by (3.21) it suffices for f to be in $H^s(\square)$.

Again, we stress that the missing assertion for an interpolator of type $\tilde{\Pi} \otimes \tilde{\Pi}^\partial$ follows analogously, even though it is not L^2-stable due to the impact of the seminorm term in (3.7). One merely needs to include the stability estimate (3.7) into either (3.16) to (3.19) or (3.20) and note that the seminorm term scales with h, and thus can be treated as part of the constant.

For an investigation of (3.13), it suffices to utilise Lemma 3.20 together with the above to see that, for $1 \leq s \leq p$, one finds

$$
\begin{aligned}
\left\|\boldsymbol{f}_1 - \tilde{\boldsymbol{\Pi}}^1_{\boldsymbol{p},\boldsymbol{\Xi}}\boldsymbol{f}_1\right\|_{\boldsymbol{H}^0(\mathrm{div},\square)} &\leq \left\|\boldsymbol{f}_1 - \tilde{\boldsymbol{\Pi}}^1_{\boldsymbol{p},\boldsymbol{\Xi}}\boldsymbol{f}_1\right\|_{L^2(\square)} + \left\|\mathrm{div}(\boldsymbol{f}_1 - \tilde{\boldsymbol{\Pi}}^1_{\boldsymbol{p},\boldsymbol{\Xi}}\boldsymbol{f}_1)\right\|_{L^2(\square)} \\
&= \left\|\boldsymbol{f}_1 - \tilde{\boldsymbol{\Pi}}^1_{\boldsymbol{p},\boldsymbol{\Xi}}\boldsymbol{f}_1\right\|_{L^2(\square)} + \left\|\mathrm{div}\,\boldsymbol{f}_1 - \mathrm{div}(\tilde{\boldsymbol{\Pi}}^1_{\boldsymbol{p},\boldsymbol{\Xi}}\boldsymbol{f}_1)\right\|_{L^2(\square)} \\
&= \left\|\boldsymbol{f}_1 - \tilde{\boldsymbol{\Pi}}^1_{\boldsymbol{p},\boldsymbol{\Xi}}\boldsymbol{f}_1\right\|_{L^2(\square)} + \left\|\mathrm{div}\,\boldsymbol{f}_1 - \tilde{\boldsymbol{\Pi}}^2_{\boldsymbol{p},\boldsymbol{\Xi}}\,\mathrm{div}(\boldsymbol{f}_1)\right\|_{L^2(\square)} \\
&\lesssim h^s \left\|\boldsymbol{f}_1\right\|_{\boldsymbol{H}^s(\mathrm{div},\square)}.
\end{aligned}
$$

The result follows by properties of the geometry mapping. ■

These results are immediately applicable to two-dimensional finite element methods with a straightforward generalisation to three dimensions, see Appendix A.

Corollary 3.23 (Approximation Results for Finite Element Methods) *Let Ω be a two-dimensional domain, satisfying Assumption 3.7. Let $f_0 \in H^1(\Omega)$, $\boldsymbol{f}_1 \in \boldsymbol{H}^0(\mathrm{div},\Omega)$ and $f_2 \in L^2(\Omega)$. Then, its holds that*

$$
\begin{aligned}
\inf_{f_h \in \mathbb{S}^0_{\boldsymbol{p},\boldsymbol{\Xi}}(\Omega)} \|f_0 - f_h\|_{H^1(\Omega)} &\lesssim h^{s-1}\|f_0\|_{H^s_{\mathrm{pw}}(\Omega)}, && \text{for } 1 \leq s \leq p+1,\\
\inf_{\boldsymbol{f}_h \in \boldsymbol{\mathbb{S}}^1_{\boldsymbol{p},\boldsymbol{\Xi}}(\Omega)} \|\boldsymbol{f}_1 - \boldsymbol{f}_h\|_{\boldsymbol{H}^0(\mathrm{div},\Omega)} &\lesssim h^{s}\|\boldsymbol{f}_1\|_{\boldsymbol{H}^s_{\mathrm{pw}}(\mathrm{div},\Omega)}, && \text{for } 0 \leq s \leq p,\\
\inf_{f_h \in \mathbb{S}^2_{\boldsymbol{p},\boldsymbol{\Xi}}(\Omega)} \|f_2 - f_h\|_{L^2(\Omega)} &\lesssim h^{s}\|f_2\|_{H^s_{\mathrm{pw}}(\Omega)}, && \text{for } 0 \leq s \leq p.
\end{aligned}
$$

Proof Due to the stability of the respective orthogonal projections onto the spline spaces, given by $\mathcal{P}_1 \colon H^1(\Omega) \to \mathbb{S}^0_{\boldsymbol{p},\boldsymbol{\Xi}}(\Omega)$, $\mathcal{P}_{\mathrm{div}} \colon \boldsymbol{H}^0(\mathrm{div},\Omega) \to \boldsymbol{\mathbb{S}}^1_{\boldsymbol{p},\boldsymbol{\Xi}}(\Omega)$ and $\mathcal{P}_0 \colon L^2(\Omega) \to \mathbb{S}^2_{\boldsymbol{p},\boldsymbol{\Xi}}(\Omega)$, we immediately have the result for the minimal values of s. Repeating the same steps as in the proof of Theorem 3.22, we find the result for larger values of s and smooth choices of f_0, $\boldsymbol{f}_1$ and f_2. The assertion now follows by interpolation arguments as in Lemma 2.17 and density of smooth functions in (subspaces of) $L^2(\Omega)$. ■

A generalisation of this result to the sequence

$$
H^1(\Omega) \xrightarrow{\ \mathbf{grad}\ } \boldsymbol{H}^0(\mathbf{curl},\Omega) \xrightarrow{\ \mathbf{curl}\ } \boldsymbol{H}^0(\mathrm{div},\Omega) \xrightarrow{\ \mathrm{div}\ } L^2(\Omega)
$$

on three-dimensional volumetric domains Ω is given in Appendix A.

3.3 Approximation W.r.t. the Energy Norms of the Trace Spaces

Now, we will consider approximation properties of the spaces $\mathbb{S}^0_{\boldsymbol{p},\boldsymbol{\Xi}}(\Gamma)$, $\mathbf{S}^1_{\boldsymbol{p},\boldsymbol{\Xi}}(\Gamma)$, and $\mathbb{S}^2_{\boldsymbol{p},\boldsymbol{\Xi}}(\Gamma)$ w.r.t. the fractional Sobolev spaces $H^{1/2}(\Gamma)$, $\boldsymbol{H}^{-1/2}_{\times}(\mathrm{div}_\Gamma,\Gamma)$, and $H^{-1/2}(\Gamma)$. While for the EFIE only the case of $\boldsymbol{H}^{-1/2}_{\times}(\mathrm{div}_\Gamma,\Gamma)$ is of relevance, for boundary element methods for Laplace or Helmholtz problems the cases of $H^{\pm 1/2}(\Gamma)$ are crucial. Corresponding isogeometric boundary element methods and their implementation have been introduced and studied by my coauthors and myself in [19].

This investigation will be conducted through the study of the corresponding orthogonal projections. Due to their optimality, we know that they must achieve the same convergence rates w.r.t. h-refinement as those of Theorem 3.22. Moreover, properties of the orthogonal projections are of interest for an application in the context of partial differential equations, since an inf-sup conditions yield quasi-optimal behaviour for the approximate solution w.r.t. the orthogonal projection of the involved energy space, see Chap. 2.

We will start by utilisation of interpolation as in Theorem 2.17 and optimality of the orthogonal projection of the respective energy space to get convergence results for positive fractional spaces. This yields the following corollary.

Corollary 3.24 (Approximating $H^{1/2}(\Gamma)$ with $\mathbb{S}^0_{\boldsymbol{p},\boldsymbol{\Xi}}(\Gamma)$) *Let Assumptions 3.5 and 3.7 hold. Let $f \in H^s_{\mathrm{pw}}(\Gamma) \cap H^{1/2}(\Gamma)$ for integers $2 \le s \le p+1$, and let $\mathcal{P}_{1/2} f$ denote its $H^{1/2}(\Gamma)$-orthogonal projection onto $\mathbb{S}^0_{\boldsymbol{p},\boldsymbol{\Xi}}(\Gamma)$. It holds that*

$$\| f - \mathcal{P}_{1/2} f \|_{H^{1/2}(\Gamma)} \lesssim h^{s-1/2} \| f \|_{H^s_{\mathrm{pw}}(\Gamma)}.$$

Proof By Theorem 3.22 we know for integers s with $2 \le s \le p+1$ that

$$\left\| f - \tilde{\Pi}^0_\Gamma(f) \right\|_{H^r(\Gamma)} \lesssim h^{s-r} \| f \|_{H^s_{\mathrm{pw}}(\Gamma)},$$

for both $r \in \{0, 1\}$. Now, application of Theorem 2.17 yields

$$\left\| f - \tilde{\Pi}^0_\Gamma(f) \right\|_{H^{1/2}(\Gamma)} \lesssim h^{s-1/2} \| f \|_{H^s_{\mathrm{pw}}(\Gamma)}.$$

By optimality of the $H^{1/2}(\Gamma)$-orthogonal projection $\mathcal{P}_{1/2}$, we obtain the result. ■

Interpolation does not yield estimates in norms with negative index. Thus, to show the approximation properties of $\mathbb{S}^2_{\boldsymbol{p},\boldsymbol{\Xi}}(\Gamma)$ in $H^{-1/2}(\Gamma)$, we resort to an application of the Aubin–Nitsche Lemma [20].

Corollary 3.25 (Approximating $H^{-1/2}(\Gamma)$ with $\mathbb{S}^2_{\boldsymbol{p},\boldsymbol{\Xi}}(\Gamma)$) *Let Assumptions 3.5 and 3.7 be satisfied. Let $f \in H^{-1/2}(\Gamma) \cap H^s_{\mathrm{pw}}(\Gamma)$ for some $s \ge 0$. Let $\mathcal{P}_{-1/2}$ denote the $H^{-1/2}(\Gamma)$-orthogonal projection of f onto $\mathbb{S}^2_{\boldsymbol{p},\boldsymbol{\Xi}}(\Gamma)$. Then it holds that*

$$\|f - \mathcal{P}_{-1/2} f\|_{H^{-1/2}(\Gamma)} \lesssim h^{s+1/2} \|f\|_{H^s_{\mathrm{pw}}(\Gamma)}, \qquad 0 \le s \le p. \tag{3.22}$$

Proof Assume, for now, that $f \in L^2(\Gamma) \cap H^s_{\mathrm{pw}}(\Gamma)$. For the L^2-orthogonal projection $\mathcal{P}_0$ onto $\mathbb{S}^2_{\boldsymbol{p},\boldsymbol{\Xi}}(\Gamma)$ we can estimate

$$\begin{aligned} \|f - \mathcal{P}_0 f\|_{H^{-1/2}(\Gamma)} &:= \sup_{0 \ne v \in H^{1/2}(\Gamma)} \frac{\left|\langle f - \mathcal{P}_0 f, v\rangle_{L^2(\Gamma)}\right|}{\|v\|_{H^{1/2}(\Gamma)}} \\ &= \sup_{0 \ne v \in H^{1/2}(\Gamma)} \frac{\left|\langle f - \mathcal{P}_0 f, v - \mathcal{P}_0 v\rangle_{L^2(\Gamma)}\right|}{\|v\|_{H^{1/2}(\Gamma)}} \\ &\lesssim \|f - \mathcal{P}_0 f\|_{L^2(\Gamma)} \sup_{0 \ne v \in H^{1/2}(\Gamma)} \frac{\|v - \mathcal{P}_0 v\|_{L^2(\Gamma)}}{\|v\|_{H^{1/2}(\Gamma)}}. \end{aligned} \tag{3.23}$$

By Theorem 3.22 and density arguments, we now arrive at $\|f - \mathcal{P}_0 f\|_{H^{-1/2}(\Gamma)} \le h^{1/2+s} \|f\|_{H^s_{\mathrm{pw}}(\Gamma)}$ for $0 \le s \le p$. Replacing $\mathcal{P}_0$ by $\mathcal{P}_{-1/2}$ now yields the assertion, using Theorem 2.17, optimality of $\mathcal{P}_{-1/2}$ w.r.t. the $H^{-1/2}(\Gamma)$-error and density of regular functions in $H^{-1/2}(\Gamma)$. ∎

Remark 3.26 This result does not necessarily rely on Theorem 3.22. Since $H^{-1/2}(\Gamma)$ allows for discontinuities, it can be reproduced by application of the patchwise estimates of Corollary 3.15. This has been done by myself and my coauthors in [19].

Remark 3.27 Note that by putting global norms on the right hand side, analogues of Corollaries 3.24 and 3.25 can be shown for minimal regularities, i.e., $1/2 \le s$ in the case of the $H^{1/2}(\Gamma)$-error and $-1/2 \le s$ in the sense of the $H^{-1/2}(\Gamma)$-error by almost analogous means, cf. [21, Theorem 4.2.17]. However, these results rely on the smoothness of the geometry for the norm on the right hand side to be well-defined.

We aim for our results to be immediately applicable to the multipatch setting of isogeometric analysis, where we want to require smoothness of the geometry only patchwise.

Now, what is missing is an analysis of the approximation properties of $\mathbf{S}^1_{\boldsymbol{p},\boldsymbol{\Xi}}(\Gamma)$ in $\boldsymbol{H}^{-1/2}_{\times}(\mathrm{div}_\Gamma, \Gamma)$. For this purpose, we want to employ an argument similar to the one of Corollary 3.25. However, as will be discussed in a moment, this cannot be done with such ease as before in Corollary 3.25. To proof the approximation result, we need to generalise the technique employed by Buffa and Christiansen [22] to larger classes of basis functions. This procedure is lengthy and technical, and presented in the upcoming Sect. 3.4.

Theorem 3.28 (Approximating $\boldsymbol{H}^{-1/2}_{\times}(\mathrm{div}_\Gamma, \Gamma)$ with $\mathbf{S}^1_{\boldsymbol{p},\boldsymbol{\Xi}}(\Gamma)$) *Let Assumptions 3.5 and 3.7 be satisfied and let $\mathcal{P}_\times \boldsymbol{f}$ denote the $\boldsymbol{H}^{-1/2}_{\times}(\mathrm{div}_\Gamma, \Gamma)$-orthogonal projection of $\boldsymbol{f}$ onto $\mathbf{S}^1_{\boldsymbol{p},\boldsymbol{\Xi}}(\Gamma)$. Let $\boldsymbol{f} \in \boldsymbol{H}^s_{\times}(\mathrm{div}_\Gamma, \Gamma)$ for some $-1/2 \le s < 0$. Then one finds*

$$\|\boldsymbol{f} - \mathcal{P}_\times \boldsymbol{f}\|_{\boldsymbol{H}^{-1/2}_{\times}(\mathrm{div}_\Gamma, \Gamma)} \lesssim h^{1/2+s} \|\boldsymbol{f}\|_{\boldsymbol{H}^s_{\times}(\mathrm{div}_\Gamma, \Gamma)}.$$

Moreover, let $\boldsymbol{f} \in \boldsymbol{H}^0(\mathrm{div}_\Gamma, \Gamma)$ *and* $\boldsymbol{f}|_{\Gamma_j} \in \boldsymbol{H}^s(\mathrm{div}_\Gamma, \Gamma_j)$ *for* $0 \leq s \leq p$ *and all patches* Γ_j. *Then one finds*

$$\|\boldsymbol{f} - \mathcal{P}_\times \boldsymbol{f}\|_{\boldsymbol{H}_\times^{-1/2}(\mathrm{div}_\Gamma,\Gamma)} \lesssim h^{1/2+s} \|\boldsymbol{f}\|_{\boldsymbol{H}^s_{\mathrm{pw}}(\mathrm{div}_\Gamma,\Gamma)}.$$

Remark 3.29 Note that Corollary 3.25 and Theorem 3.28 include the classical results from boundary element theory, even though a first glance suggests otherwise. This is due to the fact, that p refers not to the degree of $\mathbb{S}^2_{\boldsymbol{p},\boldsymbol{\Xi}}(\Gamma)$ and $\mathbf{S}^1_{\boldsymbol{p},\boldsymbol{\Xi}}(\Gamma)$ respectively, but rather to the degree at the beginning of the sequence $\mathbb{S}^0_{\boldsymbol{p},\boldsymbol{\Xi}}(\Gamma) \to \mathbf{S}^1_{\boldsymbol{p},\boldsymbol{\Xi}}(\Gamma) \to \mathbb{S}^2_{\boldsymbol{p},\boldsymbol{\Xi}}(\Gamma)$. In terms of basis functions, the space $\mathbb{S}^2_{\boldsymbol{p},\boldsymbol{\Xi}}(\Gamma)$ contains splines of degree $p-1$, thus shifting the notation by 1.

3.4 Proof of the Results for the EFIE

Within this section, we provide a detailed proof of Theorem 3.28, by means of a patch by patch duality argument, similar to the one utilised to achive the estimate in $H^{-1/2}(\Gamma)$. However, one problem with a naïve patchwise argument is due to the fact, that $\boldsymbol{H}_\times^{-1/2}(\mathrm{div}_\Gamma, \Gamma)$ incorporates a (weak) continuity across the patch normals w.r.t. $\partial\Gamma_j$, see [23]. Thus, an orthogonal approximation required for an Aubin–Nitsche type argument cannot easily be localised to a single patch.

However, the problem can be overcome through a chain of technical arguments. We will proceed as follows.

1. First, we need suitable localisation mechanisms. We will introduce a normal trace local to each patch and define a patchwise version of the $\boldsymbol{H}_\times^{-1/2}(\mathrm{div}_\Gamma, \Gamma)$-norm. This, together with analysis of the respective properties, will be done in Sect. 3.4.1.
2. In Sect. 3.4.2 we then introduce a projection operator, which approximates both the flux across patches as well as the part of the field without outgoing flux separately. We will require quasi-optimality of said projector. To prove it, we need to introduce a discrete continuous right inverse of the localised normal trace.
3. The technical lemmas and propositions will then be compiled to a proof of Theorem 3.28 in Sect. 3.4.3. The proof will assume some extra regularity, such that the technical constructions discussed in the first to parts of this section can be applied. The results will then be extended to the $\boldsymbol{H}_\times^{-1/2}(\mathrm{div}_\Gamma, \Gamma)$-orthogonal projection by density arguments.

3.4.1 Localisation and Flux Across Interfaces

We will begin by defining the space K_j^s as the kernel of the local trace operator

$$\gamma_{\boldsymbol{n},j}(\boldsymbol{f})(\boldsymbol{x}_0) := \lim_{\Gamma_j \ni \tilde{\boldsymbol{x}} \to \boldsymbol{x}_0} \boldsymbol{f}(\tilde{\boldsymbol{x}}) \cdot \boldsymbol{n}_{\boldsymbol{x}_0}, \qquad \text{for almost all } \boldsymbol{x}_0 \in \partial\Gamma_j,$$

on $\boldsymbol{H}^s(\mathrm{div}_\Gamma, \Gamma_j)$, where $\boldsymbol{n}_{\boldsymbol{x}_0}$ denotes the outer unit normal w.r.t. $\partial\Gamma_j$ at $\boldsymbol{x}_0 \in \partial\Gamma_j$. Note that the definition of the local trace operator has to be understood in a weak sense. The same way we denote the kernel of $\gamma_{\boldsymbol{n},j}$ on $\mathbb{S}^1_{p,\Xi}(\Gamma_j)$ by $K_j^{\mathbb{S}}$. Note that

$$\gamma_{\boldsymbol{n}} : \boldsymbol{H}^0(\mathrm{div}, (0,1)^2) \to H^{-1/2}(\partial\square)$$

is continuous [24, Theorem 2.5]. Due to Assumption 3.7 this immediately transfers to

$$\gamma_{\boldsymbol{n},j} : \boldsymbol{H}^0(\mathrm{div}_\Gamma, \Gamma_j) \to H^{-1/2}(\partial\Gamma_j).$$

Here, $H^{-1/2}(\partial\Gamma_j)$ has to be understood as the lifted counterpart to $H^{-1/2}(\partial\square)$, in complete analogy to the definition for two-dimensional domains, cf. [25, p. 96ff]. This definition is, again, rendered valid due to Assumption 3.7.

Remark 3.30 (*Local Shifting Property*) We remark that $\gamma_{\boldsymbol{n},j}$ enjoys a local shifting property, in the sense that $\gamma_{\boldsymbol{n},j} : \boldsymbol{H}^s(\Gamma_j) \to H^{s-1/2}(\partial\Gamma_{j,i})$ is continuous for $s > 1/2$, which implies continuity of the local trace $\gamma_{\boldsymbol{n},j} : \boldsymbol{H}^s(\mathrm{div}_\Gamma, \Gamma_j) \to H^{s-1/2}(\partial\Gamma_{j,i})$. Here, $\partial\Gamma_{j,i}$ denotes one of the four sides of $\partial\Gamma_j$. This can be seen, since in the reference domain, and restricted to one side of $\partial\square$ the identity $\gamma_{\boldsymbol{n}}(\boldsymbol{f}) = \gamma_0(\boldsymbol{e}_i^\intercal \cdot \boldsymbol{f})$ holds, where $\boldsymbol{e}_i$ is either $(1,0)^\intercal$ or $(0,1)^\intercal$, depending on the side.

This allows us to utilise the canonical continuity assertions of γ_0, compare [25, Theorem 3.37]. The argument goes as follows: We extend a given $u \in H^s((0,1)^2)$ continuously to $H^s(\mathbb{R}^2)$, see [25, Theorem A.4], apply the trace operator to the boundary ∂P of a half-plane $P \simeq [0,\infty) \times \mathbb{R}$, see [25, Lemma 3.35], and restrict it onto a given edge of $\square$. Assumption 3.7 enables us to immediately transfer this argument to the physical domain.

We now proceed by reviewing two technical results.

Lemma 3.31 (Continuity Estimate, [22, Lemma 4.8]) *Fix a patch Γ_j of the geometry. Let $\zeta \in H^{-1/2}(\partial\Gamma_j)$ such that $\langle\zeta, 1\rangle_{H^{-1/2}(\partial\Gamma_j)\times H^{1/2}(\partial\Gamma_j)} = 0$ holds. Let $\xi \in \boldsymbol{H}^0(\mathrm{div}_\Gamma, \Gamma_j)$ be the solution to the problem $\langle\xi, v\rangle_{\boldsymbol{H}^0(\mathrm{div}_\Gamma,\Gamma_j)} = 0$ for all $v \in K_j^0$ with $\gamma_{\boldsymbol{n},j}(\xi) = \zeta$ onto $\partial\Gamma_j$. Then one finds*

$$\|\xi\|_{\boldsymbol{H}^{s+1/2}(\mathrm{div}_\Gamma,\Gamma_j)} \leq C_s \|\zeta\|_{H^s(\partial\Gamma_j)}$$

for both $s = -1$ and $s = -1/2$.

Lemma 3.32 (Duality Relation, [22, Lemma 4.7]) *Let $(K_j^{-1/2})'$ denote the dual of $K_j^{-1/2}$ with respect to $\boldsymbol{H}^0(\mathrm{div}_\Gamma, \Gamma_j)$. There exists an isomorphism $K_j^{1/2} \to (K_j^{-1/2})'$.*

We shall further make use of the space $\boldsymbol{H}_{\mathrm{pw}}^{-1/2}(\mathrm{div}_\Gamma, \Gamma)$ equipped with the patchwise norm

$$\sum_{0\leq j<N} \|\cdot\|_{\boldsymbol{H}^{-1/2}(\mathrm{div}_\Gamma,\Gamma_j)}. \tag{3.24}$$

Two remarks about this definition should be made.

Remark 3.33 (*Missing L^2-Regularity*) The norm (3.24) does not incorporate the global L^2-requirement, which is included in the definition of the pw-spaces as previously stated.

Remark 3.34 (*Dual Spaces on Domains with Boundary*) To avoid technicalities, the spaces of negative Sobolev index on domains with a boundary as required by (3.24) have not been defined within this thesis. Rather, the reader is referred to the book of McLean [25, p. 77]. There, two different types of dual space, denoted by $\tilde{H}^{-s}(\Omega)$ and $H^{-s}(\Omega)$ for Ω with boundary $\partial\Omega \neq \emptyset$ and $s > 0$ are introduced. In essence, the difference between these two spaces lies in weather one considers the dual space w.r.t. functions $H_0^s(\Omega)$, i.e. the functions in $H^s(\Omega)$ whose trace onto $\partial\Omega$ vanishes, or w.r.t. the entire $H^s(\Omega)$.

For our purposes, we will consider $H^{-1/2}(\Gamma_j)$ as the dual space of $H^{1/2}(\Gamma_j)$ w.r.t. the entire $H^{1/2}(\Gamma_j)$, equipped with the corresponding norm

$$\|f\|_{H^{-1/2}(\Gamma_j)} := \sup_{g \in H^{1/2}(\Gamma_j)} \frac{\langle f, g\rangle_{L^2(\Gamma_j)}}{\|g\|_{H^{1/2}(\Gamma_j)}}.$$

The vector-valued case and the corresponding spaces with graph norm will be consider analogously.

The patchwise norm (3.24) is required to estimate the $\boldsymbol{H}_\times^{-1/2}(\mathrm{div}_\Gamma, \Gamma)$-norm locally, a fact first used in the first inequality of [22, Eq. (98)]. However, there the assertion is stated without proof. To show the corresponding estimate, we require some extra regularity, as we state a proof for functions in $\boldsymbol{H}^0(\mathrm{div}_\Gamma, \Gamma)$. There, the arguments are elementary, and the case suffices for our purposes, since the upcoming use of the quasi-interpolation operators requires even more patchwise regularity.

Proposition 3.35 (Local estimation of the $\boldsymbol{H}_\times^{-1/2}(\mathrm{div}_\Gamma, \Gamma)$*-norm*) *For* $\boldsymbol{f} \in \boldsymbol{H}^0(\mathrm{div}_\Gamma, \Gamma)$ *one finds*

$$\|\boldsymbol{f}\|_{\boldsymbol{H}_\times^{-1/2}(\mathrm{div}_\Gamma,\Gamma)} \lesssim \|\boldsymbol{f}\|_{\boldsymbol{H}_{\mathrm{pw}}^{-1/2}(\mathrm{div}_\Gamma,\Gamma)}.$$

Proof Looking at the representation (2.9), we can estimate the vector-valued term in the $\boldsymbol{H}_\times^{-1/2}(\Gamma)$-norm and the divergence term measuring the divergence in the $H^{-1/2}(\Gamma)$-norm separately. Buffa and Hiptmair [23, p. 8] state a characterisation of the $\boldsymbol{H}_\times^{1/2}(\Gamma_j)$-norm. This characterisation makes clear that $\boldsymbol{g}|_{\Gamma_j} \in \boldsymbol{H}^{1/2}(\Gamma_j)$ for any $\boldsymbol{g} \in \boldsymbol{H}_\times^{1/2}(\Gamma)$.

Moreover, from the same representation one can see that on smooth domains the spaces $\boldsymbol{H}_\times^{1/2}(\Gamma)$ and $\boldsymbol{H}^{1/2}(\Gamma)$ can be identified. This allows us to estimate any $\boldsymbol{f} \in \boldsymbol{H}_{\mathrm{pw}}^{-1/2}(\mathrm{div}_\Gamma, \Gamma)$ via

$$\begin{aligned}
\|f\|_{H_\times^{-1/2}(\Gamma)} &:= \sup_{g\in H_\times^{1/2}(\Gamma)} \frac{\langle f,g\rangle_\times}{\|g\|_{H_\times^{1/2}(\Gamma)}} \\
&= \sup_{g\in H_\times^{1/2}(\Gamma)} \sum_{0\le j<N} \frac{\langle f|_{\Gamma_j}\times n, g|_{\Gamma_j}\rangle_{L^2(\Gamma_j)}}{\|g\|_{H_\times^{1/2}(\Gamma)}} \\
&\le \sup_{g\in H_\times^{1/2}(\Gamma)} \sum_{0\le j<N} \frac{\langle f|_{\Gamma_j}\times n, g|_{\Gamma_j}\rangle_{L^2(\Gamma_j)}}{\|g|_{\Gamma_j}\|_{H^{1/2}(\Gamma_j)}} \\
&\lesssim \sum_{0\le j<N} \sup_{g_j\in H^{1/2}(\Gamma_j)} \frac{\langle f|_{\Gamma_j}\times n, g_j\rangle_{L^2(\Gamma_j)}}{\|g_j\|_{H^{1/2}(\Gamma_j)}} \\
&\lesssim \sum_{0\le j<N} \sup_{g_j\in H^{1/2}(\Gamma_j)} \frac{\langle f|_{\Gamma_j}, g_j\rangle_{L^2(\Gamma_j)}}{\|g_j\times n\|_{H^{1/2}(\Gamma_j)}} \\
&\simeq \sum_{0\le j<N} \|f|_{\Gamma_j}\|_{H^{-1/2}(\Gamma_j)}.
\end{aligned}$$

The last step uses that $(\cdot\times n)\colon H^s(\Gamma_j)\to H^s(\Gamma_j)$ is continuous for $s\in\{0,1\}$, as can be shown by straight-forward computation. Continuity for $s=1/2$ then follows from Theorem 2.17. An argument to estimate the divergence term is obtained analogously. ∎

3.4.2 A Flux Preserving Projection

We immediately define a localised projection operator that takes the kernel and preimage of $\gamma_{n,j}$ into account separately.

Definition 3.36 (*Conforming Projection*) For $g\in H^1_{\mathrm{pw}}(\Gamma)$ we define the projection $\pi=\bigoplus_{0\le j<N}\pi_j$ via patchwise projections π_j onto $\mathbb{S}^1_{p,\Xi}(\Gamma_j)$. There, each π_j is given as solution to the problem

$$\langle \pi_j g - g|_{\Gamma_j}, b\rangle_{H^0(\mathrm{div}_\Gamma,\Gamma_j)} = 0,\quad \text{for all } b\in K_j^{\mathbb{S}} \text{ and } 0\le j<N, \tag{3.25}$$

$$\langle \gamma_{n,j}(\pi_j g) - \gamma_{n,j}(g), \gamma_{n,j}(b)\rangle_{L^2(\partial\Gamma_j)} = 0,\quad \text{for all } b\in N_j^{\mathbb{S}}, \text{ and } 0\le j<N. \tag{3.26}$$

Herein, we use the decomposition

$$\mathbb{S}^1_{p,\Xi}(\Gamma_j) = N_j^{\mathbb{S}} \oplus K_j^{\mathbb{S}}, \tag{3.27}$$

where $N_j^{\mathbb{S}}$ denotes the span of basis functions with non-vanishing normal trace $\gamma_{n,j}$. Note that this induces a unique decomposition of every function in $b\in\mathbb{S}^1_{p,\Xi}(\Gamma)$, since it is clear that both $N_j^{\mathbb{S}}$ and $K_j^{\mathbb{S}}$ can be identified with specific, disjoint sets of degrees of freedom, i.e., are discrete and closed subspaces of $\mathbf{S}^1_{\boldsymbol{p},\boldsymbol{\Xi}}(\Gamma)$.

The idea behind this projection is similar to projections in the context of mixed finite element methods, which is equal to the face by face projection that preserves boundary data on interfaces, see [26]. It chooses the part without outgoing flux as the optimal approximation w.r.t. the $\boldsymbol{H}^0(\mathrm{div}_\Gamma, \Gamma_j)$-norm, and the part incorporating outgoing fluxes as optimal w.r.t. the $L^2(\partial\Gamma_j)$-norm. Since the outgoing flux is continuous across patch boundaries, (3.26) ensures the same for the discretisation.

Note that the projection is indeed well defined with respect to the composition (3.27) since each of the lines (3.25) and (3.26) fixes a unique element of $N_j^{\mathbb{S}}$ or $K_j^{\mathbb{S}}$, respectively.

Remark 3.37 (*Locality of the $L^2(\partial\Gamma_j)$-Projection*) We remark that, due to the structure of the spline space and locality of the $L^2(\partial\Gamma_j)$-scalar product, the $L^2(\partial\Gamma_j)$-orthogonal projection in (3.26) is equivalent to application of the projection to each side $\partial\Gamma_{j,i}$ of $\partial\Gamma_j$ separately.

To argue that the projector π_j has the expected approximation properties, we require a discrete right-inverse of $\gamma_{\boldsymbol{n},j}$. Such constructions are readily available, often via approximation of the continuous right-inverse. This requires for a boundary value preserving interpolation to be $\boldsymbol{H}(\mathrm{div}, \Gamma)$-stable, which is not satisfied by $\tilde{\boldsymbol{\Pi}}^1_\Gamma$. However, similar result is satisfied. By application of an inverse estimate for polynomials to (3.7) one observes that the one-dimensional interpolant $\tilde{\Pi}$ is $L^2(0, 1)$-stable for piecewise polynomial inputs. The following construction builds on this observation.

Proposition 3.38 (Discrete Right-Inverse of the Normal Trace) *There is a discrete right-inverse*

$$\boldsymbol{R}_{h,j} : \gamma_{\boldsymbol{n},j}(\mathbf{S}^1_{\boldsymbol{p},\boldsymbol{\Xi}}(\Gamma_j)) \to \mathbf{S}^1_{\boldsymbol{p},\boldsymbol{\Xi}}(\Gamma_j)$$

to $\gamma_{\boldsymbol{n},j}$ which is continuous in the sense of $H^{-1/2}(\partial\Gamma_j) \to \boldsymbol{H}^0(\mathrm{div}_\Gamma, \Gamma_j)$.

Proof The proof is simple yet technical. W.l.o.g. we conduct the argument with respect to the reference domain. First, we note that there exists a Raviart–Thomas space, that we denote by $\mathbb{Q}_{\boldsymbol{p}}$, consisting of elements $\boldsymbol{Q}_p = Q_{p,p-1} \times Q_{p-1,p}$, cf. [27] on $\square$ such that $\mathbf{S}^1_{\boldsymbol{p},\boldsymbol{\Xi}}(\square) \subseteq \mathbb{Q}_{\boldsymbol{p}}$.

For $\mathbb{Q}_{\boldsymbol{p}}$, the existence of a continuous right-inverse $\boldsymbol{R}_{\mathbb{Q}_{\boldsymbol{p}}}$ is known, see e.g. [28, Theorem 4.1.9] for lowest-order Raviart–Thomas elements. The same construction can be applied straightforwardly for high order elements, as it relies on the existence of stable quasi-interpolation operators [26, Eq. (2.5.26)], and the regularity of the continuous right-inverse, see again [28, Theorem 4.1.9] for details. Compare [27, Theorem 3.10] and Fig. 2.2, noting that in two dimensions the curl conforming spaces correspond to a rotation of the divergence conforming ones.

As a second step, we set our lifting as $\boldsymbol{R}_{h,j} = \tilde{\boldsymbol{\Pi}}^1_{\boldsymbol{p},\boldsymbol{\Xi}} \circ \boldsymbol{R}_{\mathbb{Q}_{\boldsymbol{p}}}$ and show continuity. For this, we note that the $\tilde{\Pi}$ operators commute with the surface differential operators, and that by Proposition 3.17 and the tensor-product construction (3.9) the operator $\tilde{\boldsymbol{\Pi}}^2_{\boldsymbol{p},\boldsymbol{\Xi}}$ is L^2-stable, compare also the regularity requirement (3.11). For $u \in \gamma_n(\mathbb{S}^1_{\boldsymbol{p},\boldsymbol{\Xi}}(\square)) \subseteq \gamma_n(\mathbb{Q}_p)$ we estimate

$$
\begin{aligned}
&\|\tilde{\boldsymbol{\Pi}}^1_{\boldsymbol{p},\boldsymbol{\Xi}}(\boldsymbol{R}_{\mathbb{Q}_p}u)\|^2_{\boldsymbol{H}^0(\mathrm{div},\square)}\\
&= \|\tilde{\boldsymbol{\Pi}}^1_{\boldsymbol{p},\boldsymbol{\Xi}}(\boldsymbol{R}_{\mathbb{Q}_p}u)\|^2_{L^2(\square)} + \|\mathrm{div}(\tilde{\boldsymbol{\Pi}}^1_{\boldsymbol{p},\boldsymbol{\Xi}}(\boldsymbol{R}_{\mathbb{Q}_p}u))\|^2_{L^2(\square)}\\
&= \|\tilde{\boldsymbol{\Pi}}^1_{\boldsymbol{p},\boldsymbol{\Xi}}(\boldsymbol{R}_{\mathbb{Q}_p}u)\|^2_{L^2(\square)} + \|\tilde{\boldsymbol{\Pi}}^2_{\boldsymbol{p},\boldsymbol{\Xi}}(\mathrm{div}(\boldsymbol{R}_{\mathbb{Q}_p}u))\|^2_{L^2(\square)}\\
&\lesssim \|\tilde{\boldsymbol{\Pi}}^1_{\boldsymbol{p},\boldsymbol{\Xi}}(\boldsymbol{R}_{\mathbb{Q}_p}u)\|^2_{L^2(\square)} + \|\mathrm{div}(\boldsymbol{R}_{\mathbb{Q}_p}u)\|^2_{L^2(\square)}.
\end{aligned}
$$

For the first term we estimate only one vector component, since the estimate for the second vector component follows analogously. Let R_1u denote the first vector component of $\boldsymbol{R}_{\mathbb{Q}_p}u$. Making the tensor product structure explicit, we apply the assertions of Proposition 3.17 which yields

$$
\begin{aligned}
\|(\tilde{\Pi}^{\partial}_{p,\Xi}\otimes\tilde{\Pi}_{p,\Xi})R_1u\|^2_{L^2(\square)} &\lesssim \left\|(\mathrm{Id}\otimes\tilde{\Pi}_{p,\Xi})R_1u\right\|^2_{L^2(\square)}\\
\leq \int_0^1 &\left\|(\tilde{\Pi}_{p,\Xi}\circ R_1)(u(x,\cdot))\right\|^2_{L^2(0,1)}\,\mathrm{d}x\\
\lesssim \int_0^1 &\left(\|(R_1u)(x,\cdot)\|_{L^2(0,1)} + h|(R_1u)(x,\cdot)|_{H^1(0,1)}\right)^2\,\mathrm{d}x.
\end{aligned}
$$

Note that, by choice of $\mathbb{S}^1_{\boldsymbol{p},\boldsymbol{\Xi}}(\square) \subseteq \mathbb{Q}_p$ the semi-norm term is well-defined. Since $R_1u(x,\cdot)$ is a continuous piecewise polynomial for any $x \in [0,1]$, we can apply inverse estimates [18, Lemma 4.5.3], i.e., $h|(R_1u)(x,\cdot)|_{H^1(0,1)} \lesssim \|(R_1u)(x,\cdot)\|_{L^2(0,1)}$. This together with the above yields

$$
\|\tilde{\boldsymbol{\Pi}}^1_{\boldsymbol{p},\boldsymbol{\Xi}}(\boldsymbol{R}_{\mathbb{Q}_p}u)\|_{\boldsymbol{H}^0(\mathrm{div},\square)} \lesssim \|\boldsymbol{R}_{\mathbb{Q}_p}u\|_{\boldsymbol{H}^0(\mathrm{div},\square)}.
$$

As a third step, we invoke the continuity of $\boldsymbol{R}_{\mathbb{Q}_p}$ and pull-back to the physical domain. The assertion follows. ∎

Lemma 3.39 (Convergence Property) *For any $0 \leq j < N$ and $\boldsymbol{u} \in \boldsymbol{H}^s(\mathrm{div}_\Gamma, \Gamma_j)$ the projection π_j defined in Definition 3.36 fulfils*

$$
\|\boldsymbol{u} - \pi_j\boldsymbol{u}\|_{\boldsymbol{H}^0(\mathrm{div}_\Gamma,\Gamma_j)} \lesssim h^s\|\boldsymbol{u}\|_{\boldsymbol{H}^s(\mathrm{div}_\Gamma,\Gamma_j)}, \qquad 1 \leq s \leq p.
$$

Proof Let us define the subspace of discrete functions whose normal trace coincides with $\pi_j\boldsymbol{u}$, i.e., the space

$$
\boldsymbol{W} := \{\boldsymbol{w} \in \mathbb{S}^1_{\boldsymbol{p},\boldsymbol{\Xi}}(\Gamma_j) : \gamma_{n,j}(\boldsymbol{w}) = \gamma_{n,j}(\pi_j\boldsymbol{u})\} = \{\boldsymbol{w} \in \mathbb{S}^1_{\boldsymbol{p},\boldsymbol{\Xi}}(\Gamma_j) : \boldsymbol{w} - \pi_j\boldsymbol{u} \in K^{\mathbb{S}}_j\}.
$$

In complete analogy to the proof of the Céa Lemma, see [18, Eq. (2.8.1)], we can estimate

$$
\begin{aligned}
\|\boldsymbol{u}-\pi_j\boldsymbol{u}\|^2_{\boldsymbol{H}^0(\operatorname{div}_\Gamma,\Gamma_j)} &\le \langle \boldsymbol{u}-\pi_j\boldsymbol{u},\boldsymbol{u}-\pi_j\boldsymbol{u}\rangle_{\boldsymbol{H}^0(\operatorname{div}_\Gamma,\Gamma_j)}\\
&\le \langle \boldsymbol{u}-\pi_j\boldsymbol{u},\boldsymbol{u}-\pi_j\boldsymbol{u}\rangle_{\boldsymbol{H}^0(\operatorname{div}_\Gamma,\Gamma_j)} + \underbrace{\langle \boldsymbol{u}-\pi_j\boldsymbol{u},\pi_j\boldsymbol{u}-\boldsymbol{w}_h\rangle_{\boldsymbol{H}^0(\operatorname{div}_\Gamma,\Gamma_j)}}_{\text{Orthogonal, see (3.25)}}\\
&\le \langle \boldsymbol{u}-\pi_j\boldsymbol{u},\boldsymbol{u}-\pi_j\boldsymbol{u}+\pi_j\boldsymbol{u}-\boldsymbol{w}_h\rangle_{\boldsymbol{H}^0(\operatorname{div}_\Gamma,\Gamma_j)}\\
&\le \|\boldsymbol{u}-\pi_j\boldsymbol{u}\|_{\boldsymbol{H}^0(\operatorname{div}_\Gamma,\Gamma_j)}\|\boldsymbol{u}-\boldsymbol{w}_h\|_{\boldsymbol{H}^0(\operatorname{div}_\Gamma,\Gamma_j)}
\end{aligned}
$$

for any $\boldsymbol{w}_h\in\boldsymbol{W}$, thus arriving at

$$
\left\|\boldsymbol{u}-\pi_j\boldsymbol{u}\right\|_{\boldsymbol{H}^0(\operatorname{div}_\Gamma,\Gamma_j)} \lesssim \inf_{\boldsymbol{w}\in\boldsymbol{W}} \|\boldsymbol{u}-\boldsymbol{w}\|_{\boldsymbol{H}^0(\operatorname{div}_\Gamma,\Gamma_j)}. \tag{3.28}
$$

Let us define

$$
\boldsymbol{w} = \boldsymbol{\Pi}^1_{p,\Xi}\boldsymbol{u} - \boldsymbol{R}_{h,j}\left(\gamma_{\boldsymbol{n},j}(\boldsymbol{\Pi}^1_{p,\Xi}\boldsymbol{u}) - \gamma_{\boldsymbol{n},j}(\pi_j\boldsymbol{u})\right),
$$

where $\boldsymbol{\Pi}^1_{p,\Xi}$ is the single patch operator of Lemma 3.13 lifted to Γ_j. By definition, it is immediate to see that $\gamma_{\boldsymbol{n},j}(\boldsymbol{w}) = \gamma_{\boldsymbol{n},j}(\pi_j\boldsymbol{u})$, and therefore one finds that $\boldsymbol{w}\in\boldsymbol{W}$. The triangle inequality yields

$$
\|\boldsymbol{u}-\boldsymbol{w}\|_{\boldsymbol{H}^0(\operatorname{div}_\Gamma,\Gamma_j)} \le \left\|\boldsymbol{u}-\boldsymbol{\Pi}^1_{p,\Xi}\boldsymbol{u}\right\|_{\boldsymbol{H}^0(\operatorname{div}_\Gamma,\Gamma_j)} + \left\|\boldsymbol{R}_{h,j}\left(\gamma_{\boldsymbol{n},j}(\boldsymbol{\Pi}^1_{p,\Xi}\boldsymbol{u}) - \gamma_{\boldsymbol{n},j}(\pi_j\boldsymbol{u})\right)\right\|_{\boldsymbol{H}^0(\operatorname{div}_\Gamma,\Gamma_j)}.
$$

The first term on the right can be estimated by

$$
\left\|\boldsymbol{u}-\boldsymbol{\Pi}^1_{p,\Xi}\boldsymbol{u}\right\|_{\boldsymbol{H}^0(\operatorname{div}_\Gamma,\Gamma_j)} \lesssim h^s\|\boldsymbol{u}\|_{\boldsymbol{H}^s(\operatorname{div}_\Gamma,\Gamma_j)},
$$

due to Corollary 3.15. The second term can be bounded from above by

$$
\begin{aligned}
&\left\|\boldsymbol{R}_{h,j}\left(\gamma_{\boldsymbol{n},j}(\boldsymbol{\Pi}^1_{p,\Xi}\boldsymbol{u}) - \gamma_{\boldsymbol{n},j}(\pi_j\boldsymbol{u})\right)\right\|_{\boldsymbol{H}^0(\operatorname{div}_\Gamma,\Gamma_j)}\\
&\lesssim \left\|\gamma_{\boldsymbol{n},j}(\boldsymbol{\Pi}^1_{p,\Xi}\boldsymbol{u}) - \gamma_{\boldsymbol{n},j}(\pi_j\boldsymbol{u}) + \gamma_{\boldsymbol{n},j}(\boldsymbol{u}) - \gamma_{\boldsymbol{n},j}(\boldsymbol{u})\right\|_{H^{-1/2}(\partial\Gamma_j)}\\
&\lesssim \left\|\gamma_{\boldsymbol{n},j}(\boldsymbol{\Pi}^1_{p,\Xi}\boldsymbol{u}-\boldsymbol{u})\right\|_{H^{-1/2}(\partial\Gamma_j)} + \left\|\gamma_{\boldsymbol{n},j}(\boldsymbol{u}-\pi_j\boldsymbol{u})\right\|_{H^{-1/2}(\partial\Gamma_j)}\\
&\lesssim \left\|\boldsymbol{\Pi}^1_{p,\Xi}\boldsymbol{u}-\boldsymbol{u}\right\|_{\boldsymbol{H}^0(\operatorname{div}_\Gamma,\Gamma_j)} + \left\|\gamma_{\boldsymbol{n},j}(\boldsymbol{u}) - (\mathcal{P}_0\circ\gamma_{\boldsymbol{n},j})(\boldsymbol{u})\right\|_{H^{-1/2}(\partial\Gamma_j)}.
\end{aligned}
$$

Here, we use the continuity of the lifting $\boldsymbol{R}_{h,j}$, followed by a triangle inequality, and the continuity of $\gamma_{\boldsymbol{n},j}$ together with the identity $\mathcal{P}_0\circ\gamma_{\boldsymbol{n},j} = \gamma_{\boldsymbol{n},j}\circ\pi_j$ immediately extractable from (3.26). The first term can be estimated by application of Corollary 3.15 while the second term can be handled via duality arguments. One notes that

$$
\begin{aligned}
&\left\|\gamma_{n,j}(\boldsymbol{u}) - (\mathcal{P}_0 \circ \gamma_{n,j})(\boldsymbol{u})\right\|_{H^{-1/2}(\partial\Gamma_j)} \\
&= \sup_{0\neq v\in H^{1/2}(\partial\Gamma_j)} \frac{\langle(\mathrm{Id}-\mathcal{P}_0)(\gamma_{n,j}\boldsymbol{u}), v\rangle_{L^2(\partial\Gamma_j)}}{\|v\|_{H^{1/2}(\partial\Gamma_j)}} \\
&\leq \sum_{i=0,\ldots,3}\left[\sup_{0\neq v_i\in H^{1/2}(\partial\Gamma_{j,i})} \frac{\langle(\mathrm{Id}-\mathcal{P}_0)(\gamma_{n,j}\boldsymbol{u}), v_i\rangle_{L^2(\partial\Gamma_{j,i})}}{\|v_i\|_{H^{1/2}(\partial\Gamma_{j,i})}}\right],
\end{aligned}
$$

where each $\partial\Gamma_{j,i}$ for $i = 0, \ldots 3$ corresponds to one of the smooth sides of Γ_j. In light of Remark 3.37, application of the arguments in the proof of Corollary 3.25 yields

$$
\left\|\gamma_{n,j}(\boldsymbol{u}) - (\mathcal{P}_0 \circ \gamma_{n,j})(\boldsymbol{u})\right\|_{H^{-1/2}(\partial\Gamma_j)} \lesssim h^s \sum_{i=0,\ldots,3} \left\|\gamma_{n,j}(\boldsymbol{u})\right\|_{H^{s-1/2}(\partial\Gamma_{j,i})}.
$$

One can apply the shift property of the normal trace as observed in Remark 3.30 and obtains

$$
\left\|\gamma_{n,j}(\boldsymbol{u}) - (\mathcal{P}_0 \circ \gamma_{n,j})(\boldsymbol{u})\right\|_{H^{-1/2}(\partial\Gamma_j)} \lesssim h^s \sum_{i=0,\ldots,3} \left\|\gamma_{n,j}(\boldsymbol{u})\right\|_{H^{s-1/2}(\partial\Gamma_{j,i})} \lesssim h^s \|\boldsymbol{u}\|_{\boldsymbol{H}^s(\mathrm{div}_\Gamma,\Gamma_j)}.
$$

This finally yields

$$
\|\boldsymbol{u} - \boldsymbol{w}\|_{\boldsymbol{H}^0(\mathrm{div}_\Gamma,\Gamma_j)} \lesssim h^s \|\boldsymbol{u}\|_{\boldsymbol{H}^s(\mathrm{div}_\Gamma,\Gamma_j)},
$$

which along with (3.28) concludes the proof. ∎

3.4.3 *Approximation of the Surface Current by Splines*

To complete the proof of Theorem 3.28, we need to introduce one last definition, which can be interpreted as a partial application of the conforming projection π_j.

Definition 3.40 (*Interface Approximation*) Given a function $\boldsymbol{f} \in \boldsymbol{H}^1(\Gamma_j)$, its *interface approximation* is given by $\boldsymbol{\varphi} = \boldsymbol{\varphi}_0 + \boldsymbol{\varphi}_1$, where $\boldsymbol{\varphi}_1 \in N_j^{\mathbb{S}}$ is given as the solution to the problem

$$
\langle\gamma_{n,j}(\boldsymbol{\varphi}_1), \gamma_{n,j}(\boldsymbol{b})\rangle_{L^2(\partial\Gamma_j)} = \langle\gamma_{n,j}(\boldsymbol{f}), \gamma_{n,j}(\boldsymbol{b})\rangle_{L^2(\partial\Gamma_j)}, \quad \text{for all } \boldsymbol{b} \in N_j^{\mathbb{S}}, \tag{3.29}
$$

and $\boldsymbol{\varphi}_0 \in K_j^0$ given by

$$
\langle\boldsymbol{\varphi}_0, \boldsymbol{v}\rangle_{\boldsymbol{H}^0(\mathrm{div}_\Gamma,\Gamma_j)} = \langle\boldsymbol{f} - \boldsymbol{\varphi}_1, \boldsymbol{v}\rangle_{\boldsymbol{H}^0(\mathrm{div}_\Gamma,\Gamma_j)}, \quad \text{for all } \boldsymbol{v} \in K_j^0. \tag{3.30}
$$

The interface approximation $\boldsymbol{\varphi}$ is chosen as the $\boldsymbol{H}^0(\mathrm{div}_\Gamma, \Gamma_j)$-optimal approximation of $\boldsymbol{f}$ such that the outgoing fluxes $\boldsymbol{\varphi}_1$ consist of the $L^2(\partial\Gamma_j)$-optimal approximation in the discrete sense. Note that, due to the construction of the spline space, the same is obtained if one were to apply this projection to each side of $\partial\Gamma_j$ separately. Since φ_1 as above is well-defined and the problem in (3.30) is well-posed, it is clear that φ is well defined.

Remark 3.41 We remark that we require regularity of $\boldsymbol{f}$ in Definitions 3.36 and 3.40 only for (3.26) and (3.29) to be well defined in the sense of L^2-orthogonality. Both definitions are merely technical tools to provide an estimate w.r.t. the $\boldsymbol{H}_\times^{-1/2}(\mathrm{div}_\Gamma, \Gamma)$-orthogonal projection, which, by density arguments, does not depend on the extra regularity.

We now have the required tools to show the desired convergence property. The idea of the proof is the following: Assuming regular input, we will first split the error of the approximation into two terms, one which we can estimate only w.r.t. the kernel of $\gamma_{\boldsymbol{n},j}$ via the duality relation of Lemma 3.32, and one for which we are only concerned about the outgoing flux. The second term, through the continuity of both $\gamma_{\boldsymbol{n},j}$ and the discrete right inverse, can then be reduced to a one-dimensional approximation purely on the boundary.

Estimating each term separately will yield an optimal convergence order, and, combining both terms, once again an approximation result w.r.t. a non-orthogonal projection. Replacing this by the optimal orthogonal projection allows us to retain the approximation estimate and to use density arguments to extend the estimate to non-smooth inputs.

Proof of Theorem 3.28 Fix an index $0 \leq j < N$, and assume $\boldsymbol{f}$ to be regular enough for Theorem 3.22 to be applicable. Specifically, this means that $\boldsymbol{f}$ is smooth enough for Definitions 3.36 and 3.40 to be well defined.

The triangle inequality with the interface approximation φ of $\boldsymbol{f}$ on Γ_j as intermediate element yields

$$\begin{aligned}\left\|\boldsymbol{f}|_{\Gamma_j} - \pi_j(\boldsymbol{f}|_{\Gamma_j})\right\|_{\boldsymbol{H}^{-1/2}(\mathrm{div}_\Gamma,\Gamma_j)} \leq \left\|\boldsymbol{f}|_{\Gamma_j} - \varphi\right\|_{\boldsymbol{H}^{-1/2}(\mathrm{div}_\Gamma,\Gamma_j)} \\ + \left\|\varphi - \pi_j \boldsymbol{f}\right\|_{\boldsymbol{H}^{-1/2}(\mathrm{div}_\Gamma,\Gamma_j)}.\end{aligned} \tag{3.31}$$

Let $\mathcal{P}_0$ denote the $L^2(\partial\Gamma_j)$ orthogonal projection onto $\gamma_{\boldsymbol{n},j}(\mathbb{S}^1_{\boldsymbol{p},\boldsymbol{\Xi}}(\Gamma_j))$. We will now estimate each term on the right hand side independently.

1. For the first term, we apply Lemma 3.31 with $\zeta = \gamma_{\boldsymbol{n},j}(\boldsymbol{f}) - \mathcal{P}_0(\gamma_{\boldsymbol{n},j}(\boldsymbol{f}))$ and $\xi = \boldsymbol{f}|_{\Gamma_j} - \varphi$. One readily verifies that the required assumptions are satisfied: Galerkin orthogonality yields that $\langle\zeta, 1\rangle_{H^{-1/2}(\partial\Gamma_j)\times H^{1/2}(\partial\Gamma_j)} = 0$ due to $1 \in \gamma_{\boldsymbol{n},j}\big(\mathbb{S}^1_{\boldsymbol{p},\boldsymbol{\Xi}}(\Gamma_j)\big)$ and ξ is the solution to the problem given in Lemma 3.31 due to definition of the interface approximation, see (3.30). This results in

$$\begin{aligned}\left\|\boldsymbol{f}|_{\Gamma_j} - \varphi\right\|_{\boldsymbol{H}^{-1/2}(\mathrm{div}_\Gamma,\Gamma_j)} \lesssim \left\|\gamma_{\boldsymbol{n},j}(\boldsymbol{f}) - \mathcal{P}_0(\gamma_{\boldsymbol{n},j}(\boldsymbol{f}))\right\|_{H^{-1}(\partial\Gamma_j)}, \\ \left\|\boldsymbol{f}|_{\Gamma_j} - \varphi\right\|_{\boldsymbol{H}^{0}(\mathrm{div}_\Gamma,\Gamma_j)} \lesssim \left\|\gamma_{\boldsymbol{n},j}(\boldsymbol{f}) - \mathcal{P}_0(\gamma_{\boldsymbol{n},j}(\boldsymbol{f}))\right\|_{H^{-1/2}(\partial\Gamma_j)}.\end{aligned} \tag{3.32}$$

Then, application of duality arguments as in (3.23), compare also [22, Eq. (103)], yields the estimate

$$\begin{aligned}\|f|_{\Gamma_j} - \varphi\|_{H^{-1/2}(\mathrm{div}_\Gamma,\Gamma_j)} &\lesssim \|\gamma_{n,j}(f) - \mathcal{P}_0(\gamma_{n,j}(f))\|_{H^{-1}(\partial\Gamma_j)}\\ &\lesssim h^{1/2}\|\gamma_{n,j}(f) - \mathcal{P}_0(\gamma_{n,j}(f))\|_{H^{-1/2}(\partial\Gamma_j)}\\ &\lesssim h^{1/2}\|f - \pi_j f\|_{H^0(\mathrm{div}_\Gamma,\Gamma_j)}. \end{aligned} \tag{3.33}$$

Here the last inequality is due to the continuity of the normal trace, and the fact that, by (3.26), $\mathcal{P}_0 \circ \gamma_{n,j} = \gamma_{n,j} \circ \pi_j$. Lemma 3.39 yields

$$\|f|_{\Gamma_j} - \varphi\|_{H^{-1/2}(\mathrm{div}_\Gamma,\Gamma_j)} \lesssim h^{s+1/2}\|f\|_{H^s(\mathrm{div}_\Gamma,\Gamma_j)} \tag{3.34}$$

for $1 \le s \le p$.

2. To estimate the second term of (3.31) we note that by Definition 3.40 and (3.26) we know that $\gamma_{n,j}(\varphi) = \gamma_{n,j}(\pi_j\varphi)$ holds, and thus $(\varphi - \pi_j\varphi) \in K_j^0$ follows. For all $b \in K_j^{\mathbb{S}} \subset K_j^0$ we find, through application of (3.25), (3.30) and (3.25) again, that

$$\begin{aligned}\langle \pi_j\varphi, b\rangle_{H^0(\mathrm{div}_\Gamma,\Gamma_j)} &= \langle \varphi_0 + \varphi_1, b\rangle_{H^0(\mathrm{div}_\Gamma,\Gamma_j)}\\ &= \langle f - \varphi_1 + \varphi_1, b\rangle_{H^0(\mathrm{div}_\Gamma,\Gamma_j)}\\ &= \langle \pi_j f, b\rangle_{H^0(\mathrm{div}_\Gamma,\Gamma_j)}.\end{aligned}$$

This, together with $\gamma_{n,j}(\varphi - f) = 0$ which we know since (3.26) and (3.29) coincide, implies

$$\pi_j\varphi = \pi_j f, \tag{3.35}$$

which yields $(\varphi - \pi_j f) \in K_j^0 \subset K_j^{-1/2}$. Thus, it follows that

$$\|\varphi - \pi_j f\|_{H^{-1/2}(\mathrm{div}_\Gamma,\Gamma_j)} = \sup_{0\neq v\in (K_j^{-1/2})'} \frac{\langle \varphi - \pi_j f, v\rangle_{H^0(\mathrm{div}_\Gamma,\Gamma_j)}}{\|v\|_{(K_j^{-1/2})'}}$$

holds. We stress that $K_j^0 \subset K_j^{-1/2}$ and that $K_j^{-1/2}$ is a closed subspace of $H^{-1/2}(\mathrm{div}_\Gamma, \Gamma_j)$. Lemma 3.32 and the fact that on the kernel of $\gamma_{n,j}$ the projector π_j coincides with the $H^0(\mathrm{div}_\Gamma, \Gamma_j)$-orthogonal projection onto $\mathbb{S}^1_{\Xi,p}(\Gamma)$, cf. (3.25), allow us to apply the Aubin–Nitsche technique to the above. From this follows

$$\begin{aligned}
&\sup_{0\neq \boldsymbol{v}\in (K_j^{-1/2})'} \frac{\langle \boldsymbol{\varphi} - \pi_j \boldsymbol{f}, \boldsymbol{v}\rangle_{\boldsymbol{H}^0(\mathrm{div}_\Gamma,\Gamma_j)}}{\|\boldsymbol{v}\|_{(K_j^{-1/2})'}} \\
\lesssim\ &\sup_{0\neq \boldsymbol{w}\in K_j^{1/2}} \frac{\langle \boldsymbol{\varphi} - \pi_j \boldsymbol{f}, \boldsymbol{w}\rangle_{\boldsymbol{H}^0(\mathrm{div}_\Gamma,\Gamma_j)}}{\|\boldsymbol{w}\|_{\boldsymbol{H}^{1/2}(\mathrm{div}_\Gamma,\Gamma_j)}} \\
=\ &\sup_{0\neq \boldsymbol{w}\in K_j^{1/2}} \frac{\langle \boldsymbol{\varphi} - \pi_j \boldsymbol{f}, \boldsymbol{w} - \pi_j \boldsymbol{w}\rangle_{\boldsymbol{H}^0(\mathrm{div}_\Gamma,\Gamma_j)}}{\|\boldsymbol{w}\|_{\boldsymbol{H}^{1/2}(\mathrm{div}_\Gamma,\Gamma_j)}} \\
\leq\ &\left\|\boldsymbol{\varphi} - \pi_j \boldsymbol{f}\right\|_{\boldsymbol{H}^0(\mathrm{div}_\Gamma,\Gamma_j)} \sup_{0\neq \boldsymbol{w}\in K_j^{1/2}} \frac{\left\|\boldsymbol{w} - \pi_j \boldsymbol{w}\right\|_{\boldsymbol{H}^0(\mathrm{div}_\Gamma,\Gamma_j)}}{\|\boldsymbol{w}\|_{\boldsymbol{H}^{1/2}(\mathrm{div}_\Gamma,\Gamma_j)}}.
\end{aligned}$$

Note that π_j is applicable, since (3.26) is well defined due to $\gamma_{\boldsymbol{n},j}\boldsymbol{w} = 0$ almost everywhere. Corollary 3.15 yields

$$\begin{aligned}
\left\|\boldsymbol{\varphi} - \pi_j \boldsymbol{f}\right\|_{\boldsymbol{H}^{-1/2}(\mathrm{div}_\Gamma,\Gamma_j)} &\lesssim h^{1/2}\left\|\boldsymbol{\varphi} - \pi_j \boldsymbol{f}\right\|_{\boldsymbol{H}^0(\mathrm{div}_\Gamma,\Gamma_j)} && (3.36)\\
&\lesssim h^{1/2}\left\|\boldsymbol{\varphi} - \boldsymbol{f}|_{\Gamma_j}\right\|_{\boldsymbol{H}^0(\mathrm{div}_\Gamma,\Gamma_j)} && (3.37)\\
&\quad + h^{1/2}\left\|\boldsymbol{f}|_{\Gamma_j} - \pi_j \boldsymbol{f}\right\|_{\boldsymbol{H}^0(\mathrm{div}_\Gamma,\Gamma_j)}. && (3.38)
\end{aligned}$$

The second term can again be estimated by application of Lemma 3.39. For the first term, we apply the second equation of (3.32), which we can estimate in complete analogy to (3.33), which yields

$$\left\|\boldsymbol{\varphi} - \pi_j \boldsymbol{f}\right\|_{\boldsymbol{H}^{-1/2}(\mathrm{div}_\Gamma,\Gamma_j)} \lesssim h^{s+1/2}\|\boldsymbol{f}\|_{\boldsymbol{H}^s(\mathrm{div}_\Gamma,\Gamma_j)}. \tag{3.39}$$

With both terms of (3.31) estimated by means of (3.34) and (3.39), we arrive at the patchwise estimate

$$\left\|\boldsymbol{f}|_{\Gamma_j} - \pi_j \boldsymbol{f}|_{\Gamma_j}\right\|_{\boldsymbol{H}^{-1/2}(\mathrm{div}_\Gamma,\Gamma_j)} \lesssim h^{s+1/2}\left\|\boldsymbol{f}|_{\Gamma_j}\right\|_{\boldsymbol{H}^s(\mathrm{div}_\Gamma,\Gamma_j)}.$$

Since by Proposition 3.35 the patchwise $\boldsymbol{H}_{\mathrm{pw}}^{-1/2}(\mathrm{div}_\Gamma, \Gamma)$-norm can be used to estimate the non-local $\boldsymbol{H}_\times^{-1/2}(\mathrm{div}_\Gamma, \Gamma)$-norm, we arrive at the corresponding global assertion

$$\|\boldsymbol{f} - \pi\boldsymbol{f}\|_{\boldsymbol{H}_\times^{-1/2}(\mathrm{div}_\Gamma,\Gamma)} \lesssim h^{s+1/2}\|\boldsymbol{f}\|_{\boldsymbol{H}_{\mathrm{pw}}^s(\mathrm{div}_\Gamma,\Gamma)}, \qquad 1 \leq s \leq p, \tag{3.40}$$

by properties of π. Now, stability of and optimality of $\mathcal{P}_\times$ w.r.t. $\boldsymbol{H}_\times^{-1/2}(\mathrm{div}_\Gamma, \Gamma)$ yield the estimates

$$\begin{aligned}
\|\boldsymbol{f} - \mathcal{P}_\times\boldsymbol{f}\|_{\boldsymbol{H}_\times^{-1/2}(\mathrm{div}_\Gamma,\Gamma)} &\lesssim \|\boldsymbol{f}\|_{\boldsymbol{H}_\times^{-1/2}(\mathrm{div}_\Gamma,\Gamma)}, && (3.41)\\
\|\boldsymbol{f} - \mathcal{P}_\times\boldsymbol{f}\|_{\boldsymbol{H}_\times^{-1/2}(\mathrm{div}_\Gamma,\Gamma)} &\lesssim h^{1/2}\|\boldsymbol{f}\|_{\boldsymbol{H}^0(\mathrm{div}_\Gamma,\Gamma)}, && (3.42)
\end{aligned}$$

as well as

$$\|\boldsymbol{f} - \mathcal{P}_\times \boldsymbol{f}\|_{\boldsymbol{H}_\times^{-1/2}(\mathrm{div}_\Gamma,\Gamma)} \lesssim h^{1/2}\|\boldsymbol{f}\|_{\boldsymbol{H}^0_{\mathrm{pw}}(\mathrm{div}_\Gamma,\Gamma)}, \tag{3.43}$$

$$\|\boldsymbol{f} - \mathcal{P}_\times \boldsymbol{f}\|_{\boldsymbol{H}_\times^{-1/2}(\mathrm{div}_\Gamma,\Gamma)} \lesssim h^{s+1/2}\|\boldsymbol{f}\|_{\boldsymbol{H}^s_{\mathrm{pw}}(\mathrm{div}_\Gamma,\Gamma)}. \tag{3.44}$$

By density of regular functions in $\boldsymbol{H}_\times^{-1/2}(\mathrm{div}_\Gamma, \Gamma)$ and continuity of the orthogonal projection, the results carry over to non-smooth $\boldsymbol{f}$. Now we can use interpolation to generalise the result to all $-1/2 \leq s \leq p$.[1] ■

3.5 Summary of the Approximation Properties

Up to this point, we have derived multipatch approximation results of the spline complex w.r.t. the norms required by boundary- and finite element methods.

Let the functions f_0, $\boldsymbol{f}_1$, f_2 be regular enough for the norms on both left and right-hand side of the following estimates to be well defined, see also Lemma 3.20. For boundaries Γ in accordance with Assumptions 3.5 and 3.7, we proved

$$\inf_{f_h\in\mathbb{S}^0_{\boldsymbol{p},\boldsymbol{\Xi}}(\Gamma)} \|f_0 - f_h\|_{H^{1/2}(\Gamma)} \lesssim h^{s-1/2}\|f_0\|_{H^s_{\mathrm{pw}}(\Gamma)} \qquad 2 \leq s \leq p+1, \tag{3.45}$$

$$\inf_{\boldsymbol{f}_h\in\mathbb{S}^1_{\boldsymbol{p},\boldsymbol{\Xi}}(\Gamma)} \|\boldsymbol{f}_1 - \boldsymbol{f}_h\|_{\boldsymbol{H}_\times^{-1/2}(\mathrm{div}_\Gamma,\Gamma)} \lesssim h^{s+1/2}\|\boldsymbol{f}_1\|_{\boldsymbol{H}^s_\times(\mathrm{div}_\Gamma,\Gamma)} \qquad -1/2 \leq s \leq 0, \tag{3.46}$$

$$\inf_{\boldsymbol{f}_h\in\mathbb{S}^1_{\boldsymbol{p},\boldsymbol{\Xi}}(\Gamma)} \|\boldsymbol{f}_1 - \boldsymbol{f}_h\|_{\boldsymbol{H}_\times^{-1/2}(\mathrm{div}_\Gamma,\Gamma)} \lesssim h^{s+1/2}\|\boldsymbol{f}_1\|_{\boldsymbol{H}^s_{\mathrm{pw}}(\mathrm{div}_\Gamma,\Gamma)} \qquad 0 \leq s \leq p, \tag{3.47}$$

$$\inf_{f_h\in\mathbb{S}^2_{\boldsymbol{p},\boldsymbol{\Xi}}(\Gamma)} \|f_2 - f_h\|_{H^{-1/2}(\Gamma)} \lesssim h^{s+1/2}\|f_2\|_{H^s_{\mathrm{pw}}(\Gamma)} \qquad 0 \leq s \leq p. \tag{3.48}$$

Here, (3.45) follows from Corollary 3.24, (3.46) and (3.47) follow from Theorem 3.28, and (3.48) follows from Corollary 3.25. Moreover, we can apply these results for finite element methods as well. We find for multipatch domains $\Omega \subseteq \mathbb{R}^2$ the estimates

$$\inf_{f_h\in\mathbb{S}^0_{\boldsymbol{p},\boldsymbol{\Xi}}(\Omega)} \|f_3 - f_h\|_{H^1(\Omega)} \lesssim h^{s-1}\|f_3\|_{H^s_{\mathrm{pw}}(\Omega)} \qquad 2 \leq s \leq p+1,$$

$$\inf_{\boldsymbol{f}_h\in\mathbb{S}^1_{\boldsymbol{p},\boldsymbol{\Xi}}(\Omega)} \|\boldsymbol{f}_4 - \boldsymbol{f}_h\|_{\boldsymbol{H}^0(\mathrm{div},\Omega)} \lesssim h^{s}\|\boldsymbol{f}_4\|_{\boldsymbol{H}^s_{\mathrm{pw}}(\mathrm{div},\Omega)} \qquad 1 \leq s \leq p,$$

$$\inf_{f_h\in\mathbb{S}^2_{\boldsymbol{p},\boldsymbol{\Xi}}(\Omega)} \|f_5 - f_h\|_{L^2(\Omega)} \lesssim h^{s}\|f_5\|_{H^s_{\mathrm{pw}}(\Omega)} \qquad 0 \leq s \leq p,$$

[1] This can be done thanks to Appendix 2 of [22], which proves that $\boldsymbol{H}_\times^{-1/2}(\mathrm{div}_\Gamma, \Gamma)$ and $\boldsymbol{H}^0(\mathrm{div}_\Gamma, \Gamma)$ induce an interpolation scale, i.e., can be handled similarly to Theorem 2.17. Specifically, see [22, Theorem 4.12] where the notation translates to ours via $X = \boldsymbol{H}_\times^{-1/2}(\mathrm{div}_\Gamma, \Gamma)$ and $X^s = \boldsymbol{H}^s_\times(\mathrm{div}_\Gamma, \Gamma)$.

for $f_3, \boldsymbol{f}_4$ and f_5 smooth enough for the norms to be defined, as explained in Corollary 3.23. By extension of the tensor product structure in the construction of spline spaces and interpolation operators by one dimension, the given construction can easily be generalised, see Appendix A. Estimates for three-dimensional spaces, including $\boldsymbol{H}(\mathbf{curl}, \Omega)$, follow analogously, cf. Corollary A.1 in Appendix A. We can drop the regularity requirements from Theorem 3.22, since they are only required by the constructed quasi-interpolants, and not by the orthogonal projection w.r.t. the corresponding Sobolev spaces, see Sect. 3.3.

Taking into account the three-dimensional generalisation of the construction, we now have access to a discretisation of the diagram in Fig. 2.2, given by

$$\begin{array}{ccccccc}
\mathbb{S}^0_{\tilde{\boldsymbol{p}},\tilde{\boldsymbol{\Xi}}}(\Omega) & -\mathbf{grad}\rightarrow & \boldsymbol{\mathbb{S}}^1_{\tilde{\boldsymbol{p}},\tilde{\boldsymbol{\Xi}}}(\Omega) & -\mathbf{curl}\rightarrow & \boldsymbol{\mathbb{S}}^2_{\tilde{\boldsymbol{p}},\tilde{\boldsymbol{\Xi}}}(\Omega) & -\mathrm{div}\rightarrow & \mathbb{S}^3_{\tilde{\boldsymbol{p}},\tilde{\boldsymbol{\Xi}}}(\Omega) \\
\downarrow \gamma_0 & & \downarrow \boldsymbol{\gamma}_t & & \downarrow \gamma_{\boldsymbol{n}} & & \\
\mathbb{S}^0_{\boldsymbol{p},\boldsymbol{\Xi}}(\Gamma) & -\mathbf{curl}_\Gamma\rightarrow & \boldsymbol{\mathbb{S}}^1_{\boldsymbol{p},\boldsymbol{\Xi}}(\Gamma) & -\mathrm{div}_\Gamma\rightarrow & \mathbb{S}^2_{\boldsymbol{p},\boldsymbol{\Xi}}(\Gamma) & &
\end{array} \tag{3.49}$$

for suitable choices of (lists of tuples of) polynomial degrees $\tilde{\boldsymbol{p}}$, $\boldsymbol{p}$ and knot vectors $\tilde{\boldsymbol{\Xi}}$, $\boldsymbol{\Xi}$, a multipatch domain Ω. Note that a corresponding discretisation of the rotated counterpart $\boldsymbol{H}^{-1/2}(\mathrm{curl}_\Gamma, \Gamma)$ can be obtained in complete analogy to the construction of $\boldsymbol{\mathbb{S}}^1_{\boldsymbol{p},\boldsymbol{\Xi}}(\Gamma)$.

To this end, we know that for *any* problem formulated within the isogeometric framework that enjoys a discrete inf-sup condition as in Theorem 2.11 or allow for an application of Céa's Lemma as in Theorem 2.9 w.r.t. the norms above, we can expect a convergence of optimal order w.r.t. h-refinement [29]. Note, however, that the orthogonal projection will, in general, not have the commuting diagram property in the sense of Lemma 3.20. This distinction is critical for existence and uniqueness proofs for problems requiring conforming discretisations, as will become clear in the remainder of this chapter.

3.6 The Isogeometric Electric Field Integral Equation

This section will now apply the proven approximation properties to the electric field integral equation, results first published in [6]. To this end, replacing $\boldsymbol{H}_\times^{-1/2}(\mathrm{div}_\Gamma, \Gamma)$ by $\boldsymbol{\mathbb{S}}^1_{\boldsymbol{p},\boldsymbol{\Xi}}(\Gamma) \subset \boldsymbol{H}_\times^{-1/2}(\mathrm{div}_\Gamma, \Gamma)$ yields the discrete variational problem to Problem 2.37, given as follows.

Problem 3.42 (*Discrete Problem*) Find $\boldsymbol{w}_h \in \boldsymbol{\mathbb{S}}^1_{\boldsymbol{p},\boldsymbol{\Xi}}(\Gamma)$ such that

$$\langle \mathbf{V}_\kappa \boldsymbol{w}_h, \boldsymbol{\mu}_h \rangle_\times = -\langle \boldsymbol{\gamma}_t \boldsymbol{e}_i, \boldsymbol{\mu}_h \rangle_\times, \tag{3.50}$$

for all $\boldsymbol{\mu}_h \in \boldsymbol{\mathbb{S}}^1_{\boldsymbol{p},\boldsymbol{\Xi}}(\Gamma)$.

Given a basis $\{\varphi_i\}_{i=1}^N$ of $\mathbb{S}^1_{\boldsymbol{p},\boldsymbol{\Xi}}(\Gamma)$, this yields the linear system

$$\mathbf{V}_{\kappa,h}\boldsymbol{w}_h = -\boldsymbol{f}_h, \tag{3.51}$$

where the right-hand side $\boldsymbol{f}_h$ is given by $\left[\boldsymbol{f}_h\right]_j = \langle \gamma_t \boldsymbol{e}_i, \boldsymbol{\varphi}_j\rangle_\times$ and the system matrix $\mathbf{V}_{\kappa,h}$ by

$$\begin{aligned}\left[\mathbf{V}_{\kappa,h}\right]_{i,j} &= \langle \mathcal{V}_\kappa \boldsymbol{\varphi}_j, \boldsymbol{\varphi}_i\rangle_\times \\ &= \int_\Gamma\int_\Gamma u^*_\kappa(\boldsymbol{x}-\boldsymbol{y})\boldsymbol{\varphi}_j(\boldsymbol{x})\cdot\boldsymbol{\varphi}_i(\boldsymbol{y})\,\mathrm{d}\sigma_{\boldsymbol{y}}\,\mathrm{d}\sigma_{\boldsymbol{x}} \\ &\quad -\frac{1}{\kappa^2}\int_\Gamma\int_\Gamma u^*_\kappa(\boldsymbol{x}-\boldsymbol{y})\,\mathrm{div}_\Gamma\,\boldsymbol{\varphi}_j(\boldsymbol{x})\,\mathrm{div}_\Gamma\,\boldsymbol{\varphi}_i(\boldsymbol{y})\,\mathrm{d}\sigma_{\boldsymbol{y}}\,\mathrm{d}\sigma_{\boldsymbol{x}},\end{aligned} \tag{3.52}$$

see [22, Eq. (32)]. We remark that the system matrix is symmetric, but not Hermitian.

According to the classical theory of the electric field integral equation the following holds.

Lemma 3.43 (Criteria for a Stable Discretisation, [23, Proposition 4.1]) *Assume the following.*

1. *There exists a continuous splitting* $\boldsymbol{H}^{-1/2}_\times(\mathrm{div}_\Gamma,\Gamma) = \boldsymbol{W}\oplus\boldsymbol{V}$ *such that the bilinear form induced by Problem 2.37 is stable and coercive on* $\boldsymbol{V}\times\boldsymbol{V}$ *and* $\boldsymbol{W}\times\boldsymbol{W}$, *and compact on* $\boldsymbol{V}\times\boldsymbol{W}$ *and* $\boldsymbol{W}\times\boldsymbol{V}$.
2. *The space* $\mathbb{S}^1_{\boldsymbol{p},\boldsymbol{\Xi}}(\Gamma)$ *can be decomposed into a sum*

$$\mathbb{S}^1_{\boldsymbol{p},\boldsymbol{\Xi}}(\Gamma) := \boldsymbol{W}_h \oplus \boldsymbol{V}_h$$

of closed subspaces of $\boldsymbol{H}^{-1/2}_\times(\mathrm{div}_\Gamma,\Gamma)$.
3. $\boldsymbol{W}_h$ *and* $\boldsymbol{V}_h$ *are stable under complex conjugation, and*
4. *One finds* $\boldsymbol{W}_h\subseteq\boldsymbol{W}$, *and the so-called* gap-property

$$\sup_{\boldsymbol{v}_h\in\boldsymbol{V}_h}\inf_{\boldsymbol{v}\in\boldsymbol{V}}\frac{\|\boldsymbol{v}-\boldsymbol{v}_h\|_{\boldsymbol{H}^{-1/2}_\times(\mathrm{div}_\Gamma,\Gamma)}}{\|\boldsymbol{v}_h\|_{\boldsymbol{H}^{-1/2}_\times(\mathrm{div}_\Gamma,\Gamma)}} \overset{h\to 0}{\longrightarrow} 0. \tag{3.53}$$

holds.

Then Problem 2.31 satisfies the assumptions of Theorem 2.11, i.e., is inf-sup-*stable.*

The continuous splitting of $\boldsymbol{H}^{-1/2}_\times(\mathrm{div}_\Gamma,\Gamma)$ has been discussed for non-resonant wavenumbers by, e.g., Buffa and Christiansen [22], and is required for proving the inf-sup-stability of Problem 2.37. One of the most concise (although not self-contained) constructions of said splitting and the discrete inf-sup condition according to this scheme is presented by Bespalov et al. [30], whose lines we will follow closely, starting with the introduction of some necessary operators. The theory behind them goes back to Hiptmair [31].

Lemma 3.44 (Regularising Projection, [30, Lemma 3.1.]) *For compact domains* Ω *with Lipschitz boundary there exists a continuous projection* $\mathsf{R}\colon \boldsymbol{H}_{\times}^{-1/2}(\mathrm{div}_\Gamma, \Gamma) \to \boldsymbol{H}_{\times}^{1/2}(\Gamma)$ *such that*

$$(\mathrm{div}_\Gamma \circ \mathsf{R})(\boldsymbol{u}) = \mathrm{div}_\Gamma \boldsymbol{u}, \tag{3.54}$$

for all $\boldsymbol{u} \in \boldsymbol{H}_{\times}^{-1/2}(\mathrm{div}_\Gamma, \Gamma)$.

Proof We shortly review the proof for later reference. For any $\boldsymbol{v} \in \boldsymbol{H}_{\times}^{-1/2}(\mathrm{div}_\Gamma, \Gamma)$, the solution of the Neumann problem

$$\begin{aligned} \Delta w &= 0 && \text{in } \Omega, \\ \mathbf{grad}\, w \cdot \boldsymbol{n} &= \mathrm{div}_\Gamma \boldsymbol{v} && \text{on } \Gamma, \end{aligned} \tag{3.55}$$

defines a vector field $\mathbf{grad}\, w \in \boldsymbol{H}^0(\mathrm{div}\, 0, \Omega)$ with $\langle \mathbf{grad}\, w \cdot \boldsymbol{n}, 1\rangle_{L^2(\Gamma)} = 0$. Herein, $\boldsymbol{H}^0(\mathrm{div}\, 0, \Omega)$ denotes the divergence-free subspace of $L^2(\Omega)$. Using a continuous lifting operator $\mathsf{L}\colon \boldsymbol{H}^0(\mathrm{div}, \Omega) \to \boldsymbol{H}^1(\Omega)$ with $\mathbf{curl}\mathsf{L}\boldsymbol{u} = \boldsymbol{u}$ for all $\boldsymbol{u} \in \boldsymbol{H}^0(\mathrm{div}\, 0, \Omega)$ satisfying $\langle \boldsymbol{u} \cdot \boldsymbol{n}, 1\rangle_{L^2(\Gamma)} = 0$, see, e.g., [24, Theorem 3.4], we finally arrive at $\mathsf{R}\boldsymbol{v} := \gamma_t \mathsf{L}\mathbf{grad}\, w \in \boldsymbol{H}_{\times}^{1/2}(\Gamma)$. Property (3.54) follows. Since $\mathbf{grad}\, w$ depends continuously on $\mathrm{div}_\Gamma \boldsymbol{v}$ and $\gamma_t\colon \boldsymbol{H}^1(\Omega) \to \boldsymbol{H}_{\times}^{1/2}(\Gamma)$ is continuous as well, the continuity of R follows. ∎

Indeed, one can show the continuous inf-sup-condition via the splitting $\boldsymbol{V} := \mathsf{R}\big(\boldsymbol{H}_{\times}^{-1/2}(\mathrm{div}_\Gamma, \Gamma)\big)$ and $\boldsymbol{W} := (\mathrm{Id} - \mathsf{R})\big(\boldsymbol{H}_{\times}^{-1/2}(\mathrm{div}_\Gamma, \Gamma)\big)$. However, in contrast to Bespalov et al. [30], we cannot use the operator R to introduce a discrete splitting, since the image of R is not patch-wise in $\boldsymbol{H}^1$, which would be required for a function to be contained in the domain of $\tilde{\boldsymbol{\Pi}}_\Gamma^1$. Instead, we have to introduce another regularising projection.

Lemma 3.45 (Regularising Projection of Higher Regularity) *For compact* Ω *with Lipschitz boundary there exists a continuous projection*

$$\mathsf{R}_0\colon \boldsymbol{H}^0(\mathrm{div}_\Gamma, \Gamma) \to \boldsymbol{H}_{\mathrm{pw}}^1(\Gamma)$$

such that

$$(\mathrm{div}_\Gamma \circ \mathsf{R}_0)(\boldsymbol{u}) = \mathrm{div}_\Gamma \boldsymbol{u} \tag{3.56}$$

holds for all $\boldsymbol{u} \in \boldsymbol{H}^0(\mathrm{div}_\Gamma, \Gamma)$.

Proof The proof is in analogy to Lemma 3.44. First, we remark that $\mathrm{div}_\Gamma \boldsymbol{u} \in L^2(\Gamma)$. Thus, (3.55) yields a field $\mathbf{grad}\, w \in \boldsymbol{H}^{1/2}(\mathrm{div}\, 0, \Omega)$ with

$$\langle \mathbf{grad}\, w \cdot \boldsymbol{n}, 1\rangle_{L^2(\Gamma)} = 0.$$

Girault and Raviart [24, Remark 3.12] show that there is a continuous lifting

$$\mathsf{L}_{1/2} \colon \boldsymbol{H}^{1/2}(\operatorname{div} 0, \Omega) \to \boldsymbol{H}^{3/2}(\Omega)$$

with $\mathbf{curl}\mathsf{L}\boldsymbol{u} = \boldsymbol{u}$ for all $\boldsymbol{u} \in \boldsymbol{H}^{1/2}(\operatorname{div} 0, \Omega)$ satisfying $\langle \boldsymbol{u} \cdot \boldsymbol{n}, 1\rangle_{L^2(\Gamma)} = 0$. This yields the assertion by patchwise application of the rotated tangential trace. We remark that the continuity of $\mathsf{L}_{1/2}$ follows by noting that the construction of the extensions in [24, Theorem 3.4] and [24, Corollary 3.3] depend continuously on the input data. The interpolation argument of [24, Remark 3.12] then yields the continuity assertion, since the image of the procedure in [24, Theorem 3.4] and [24, Corollary 3.3] coincides for equivalent input data. ■

In analogy to the continuous setting, by Lemma 3.45 and the construction and properties of the quasi-interpolation operators constructed in [5], the definition of the discrete splitting via

$$\boldsymbol{V}_h := (\tilde{\boldsymbol{\Pi}}_\Gamma^1 \circ \mathsf{R}_0)\big(\mathbb{S}^1_{\boldsymbol{p},\boldsymbol{\Xi}}(\Gamma)\big), \qquad \boldsymbol{W}_h := (\operatorname{Id} - \tilde{\boldsymbol{\Pi}}_\Gamma^1 \circ \mathsf{R}_0)\big(\mathbb{S}^1_{\boldsymbol{p},\boldsymbol{\Xi}}(\Gamma)\big),$$

is well-defined and will prove to be a suitable candidate to fulfill the assumptions of Lemma 3.43.

Remark 3.46 The construction of both R and R_0 shows that the kernel of the respective operator consists exactly of the divergence-free functions. Thus, it follows that $\boldsymbol{W}_h \subseteq \boldsymbol{W}$ holds, compare [30, Eq. 3.5].

We are now ready to provide a statement about the inf-sup-stability of the discretised EFIE.

Theorem 3.47 (Discrete Inf-Sup Condition) *For non-resonant wavenumbers problem 3.42 is* inf-sup-*stable in the sense of Theorem 2.11.*

Proof First, we consider the case of where $\mathbb{S}^1_{\boldsymbol{p},\boldsymbol{\Xi}}(\Gamma)$ is patchwise continuous. Due to Remark 3.46, it remains to check the gap property (3.53) for $\boldsymbol{V}$ and $\boldsymbol{V}_h$. We write

$$\begin{aligned}
\sup_{\boldsymbol{v}_h \in \boldsymbol{V}_h} \inf_{\boldsymbol{v} \in \boldsymbol{V}} \frac{\|\boldsymbol{v} - \boldsymbol{v}_h\|_{\boldsymbol{H}_\times^{-1/2}(\operatorname{div}_\Gamma,\Gamma)}}{\|\boldsymbol{v}_h\|_{\boldsymbol{H}_\times^{-1/2}(\operatorname{div}_\Gamma,\Gamma)}} &\lesssim \sup_{\boldsymbol{v}_h \in \boldsymbol{V}_h} \frac{\|\mathsf{R}_0\boldsymbol{v}_h - \boldsymbol{v}_h\|_{\boldsymbol{H}_\times^{-1/2}(\operatorname{div}_\Gamma,\Gamma)}}{\|\boldsymbol{v}_h\|_{\boldsymbol{H}_\times^{-1/2}(\operatorname{div}_\Gamma,\Gamma)}} \\
&= \sup_{\boldsymbol{v}_h \in \boldsymbol{V}_h} \frac{\left\|(\mathsf{R}_0 - \tilde{\boldsymbol{\Pi}}_\Gamma^1 \circ \mathsf{R}_0)(\boldsymbol{v}_h)\right\|_{\boldsymbol{H}_\times^{-1/2}(\operatorname{div}_\Gamma,\Gamma)}}{\|\boldsymbol{v}_h\|_{\boldsymbol{H}_\times^{-1/2}(\operatorname{div}_\Gamma,\Gamma)}}.
\end{aligned}$$

Note that the last equality holds because one can show that $\tilde{\boldsymbol{\Pi}}_\Gamma^1 \circ \mathsf{R}_0$ is a projection, as done by Bespalov et al. [30, Sect. 6] for R. Thus, it holds that

$$(\tilde{\boldsymbol{\Pi}}_\Gamma^1 \circ \mathsf{R}_0)(\boldsymbol{V}_h) = \boldsymbol{V}_h := (\tilde{\boldsymbol{\Pi}}_\Gamma^1 \circ \mathsf{R}_0)(\mathbb{S}^1_{\boldsymbol{p},\boldsymbol{\Xi}}(\Gamma)).$$

Since the embedding $\boldsymbol{H}^0(\operatorname{div}_\Gamma, \Gamma) \to \boldsymbol{H}_\times^{-1/2}(\operatorname{div}_\Gamma, \Gamma)$ is continuous we arrive at

$$\left\|(\mathsf{R}_0 - \tilde{\Pi}^1_\Gamma \circ \mathsf{R}_0)(\boldsymbol{v}_h)\right\|_{\boldsymbol{H}_\times^{-1/2}(\mathrm{div}_\Gamma,\Gamma)} \lesssim \left\|(\mathsf{R}_0 - \tilde{\Pi}^1_\Gamma \circ \mathsf{R}_0)(\boldsymbol{v}_h)\right\|_{\boldsymbol{H}^0(\mathrm{div}_\Gamma,\Gamma)}.$$

By the divergence preserving property (3.56) of R_0 and the fact that the interpolation operators are projections which commute w.r.t. the surface differential operator, we can apply Theorem 3.22. Thus we arrive at

$$\left\|(\mathsf{R}_0 - \tilde{\Pi}^1_\Gamma \circ \mathsf{R}_0)(\boldsymbol{v}_h)\right\|_{\boldsymbol{H}^0(\mathrm{div}_\Gamma,\Gamma)} = \left\|(\mathsf{R}_0 - \tilde{\Pi}^1_\Gamma \circ \mathsf{R}_0)(\boldsymbol{v}_h)\right\|_{\boldsymbol{L}^2(\Gamma)} \lesssim h\|\mathsf{R}_0\boldsymbol{v}_h\|_{\boldsymbol{H}^1_{\mathrm{pw}}(\Gamma)}.$$

Note that the right hand side is well-defined due to $\mathsf{R}_0(\boldsymbol{V}_h) \subset \boldsymbol{H}^1_{\mathrm{pw}}(\Gamma)$. Combining the above with the continuity of the R_0 operator and an inverse estimate, provided in [6, Lemma A.1] for patchwise continuous basis functions, yields

$$\left\|(\mathsf{R}_0 - \tilde{\Pi}^1_\Gamma \circ \mathsf{R}_0)(\boldsymbol{v}_h)\right\|_{\boldsymbol{H}_\times^{-1/2}(\mathrm{div}_\Gamma,\Gamma)} \lesssim h^{1/2}\|\boldsymbol{v}_h\|_{\boldsymbol{H}_\times^{-1/2}(\mathrm{div}_\Gamma,\Gamma)}.$$

Thus the assertion for the patchwise continuous case. The case where $\mathbb{S}^1_{\boldsymbol{p},\boldsymbol{\Xi}}(\Gamma)$ is not continuous within each patch is realised only for maximal knot repetition within knot vectors. It reduces either to analogous arguments applied to subdomains satisfying the continuity assumption, or, otherwise reduces to the classical theory of higher order Raviart–Thomas elements on quadrilaterals for which we refer to the works of Buffa and Hiptmair [23] and Zaglmayr [32]. Compare also the proof of Proposition 3.38, which uses a similar approach. ∎

Following classical theory as established by Babuška [33] we can finally combine Theorems 3.28 and 3.47 and arrive at the main result of this subsection.

Theorem 3.48 (Discretisation Error) *For non-resonant wavenumbers the solution to Problem 3.42 exists and is unique. Moreover, assuming $\boldsymbol{w} \in \boldsymbol{H}^s_{\mathrm{pw}}(\mathrm{div}_\Gamma,\Gamma)$ for some $0 < s \le p$, for the solutions $\boldsymbol{w} \in \boldsymbol{H}_\times^{-1/2}(\mathrm{div}_\Gamma,\Gamma)$ and $\boldsymbol{w}_h \in \mathbb{S}^1_{\boldsymbol{p},\boldsymbol{\Xi}}(\Gamma)$ of Problem 2.37 and Problem 3.42 we find that*

$$\|\boldsymbol{w} - \boldsymbol{w}_h\|_{\boldsymbol{H}_\times^{-1/2}(\mathrm{div}_\Gamma,\Gamma)} \lesssim h^{s+1/2}\|\boldsymbol{w}\|_{\boldsymbol{H}^s_{\mathrm{pw}}(\mathrm{div}_\Gamma,\Gamma)}.$$

As a corollary, we can predict the expected convergence rates of the scattered electric field. Similar to scalar-valued problems, the convergence rate of the field doubles.

Corollary 3.49 (Pointwise Error) *Let $\boldsymbol{x} \in \Omega^c$ fixed. Let $\boldsymbol{w}$ be the solution to Problem 2.37 and $\boldsymbol{w}_h$ the solution to Problem 3.42. Then, for $\boldsymbol{E}_s = \mathcal{V}_\kappa\boldsymbol{w}$ and $\boldsymbol{e}_{s,h} = \mathcal{V}_\kappa\boldsymbol{w}_h$, it holds*

$$\|\boldsymbol{E}_s(\boldsymbol{x}) - \boldsymbol{E}_{s,h}(\boldsymbol{x})\|_{\mathbb{C}^3} \le C(\boldsymbol{x})h^{2p+1}\|\boldsymbol{w}\|_{\boldsymbol{H}^p_{\mathrm{pw}}(\mathrm{div}_\Gamma,\Gamma)},$$

if $\boldsymbol{w}$ and the solution of a suitable adjoint problem are sufficiently smooth.

Proof One readily verifies that $(\mathcal{V}_\kappa \cdot)(\boldsymbol{x})$ is a linear and continuous functional on $\boldsymbol{H}_\times^{-1/2}(\mathrm{div}_\Gamma, \Gamma)$ for given $\boldsymbol{x}$. Let then $\boldsymbol{\varphi}^{(\boldsymbol{x})}$ be the solution of the adjoint problem of finding $\boldsymbol{\varphi}^{(x)} \in \boldsymbol{H}_\times^{-1/2}(\mathrm{div}_\Gamma, \Gamma)$ such that

$$\langle \mathrm{V}_\kappa \boldsymbol{\xi}, \boldsymbol{\varphi}^{(x)} \rangle_\times = (\mathcal{V}_\kappa \boldsymbol{\xi})(\boldsymbol{x}) \tag{3.57}$$

holds for all $\boldsymbol{\xi} \in \boldsymbol{H}_\times^{-1/2}(\mathrm{div}_\Gamma, \Gamma)$. Let $\boldsymbol{\varphi}_h^{(\boldsymbol{x})}$ denote its discrete analogon. The assertion now follows by applying [21, Theorem 4.2.14] to each component of the scattered field to obtain

$$\|\boldsymbol{E}_s(\boldsymbol{x}) - \boldsymbol{E}_{s,h}(\boldsymbol{x})\|_{\mathbb{C}^3} \lesssim \|\boldsymbol{w} - \boldsymbol{w}_h\|_{\boldsymbol{H}_\times^{-1/2}(\mathrm{div}_\Gamma,\Gamma)} \left\|\boldsymbol{\varphi}^{(\boldsymbol{x})} - \boldsymbol{\varphi}_h^{(\boldsymbol{x})}\right\|_{\boldsymbol{H}_\times^{-1/2}(\mathrm{div}_\Gamma,\Gamma)}.$$

The previous theorem yields the assertion with $C(\boldsymbol{x}) = C\|\boldsymbol{\varphi}^{(\boldsymbol{x})}\|_{\boldsymbol{H}^p_{\mathrm{pw}}(\mathrm{div}_\Gamma,\Gamma)}$, if the solutions to Problem 2.37 and (3.57) are smooth enough. ■

Remark 3.50 The proof also applies to linear and continuous output functionals of $\boldsymbol{w}$. Thus, similar error estimates hold also for other quantities of interest, for example for path integrals of the electric field, i.e., voltages, radar cross sections, see the book of Jackson [34], or even the solutions to the corresponding eigenvalue problems, as will be discussed later in this thesis.

References

1. Cottrell JA, Hughes TJR, Bazilevs Y (2009) Isogeometric analysis: toward integration of CAD and FEA. Wiley, New York
2. Hughes TJR, Cottrell JA, Bazilevs Y (2005) Isogeometric analysis: CAD, finite elements, NURBS, exact geometry and mesh refinement. Comput Methods Appl Mech Eng 194:4135–4195
3. Piegl L, Tiller W (1987) Curve and surface constructions using rational B-splines. Comput-Aided Design 19(9):485–498
4. Schumaker LL (2007) Spline functions: basic theory (English). Cambridge mathematical library. Cambridge University Press, Cambridge
5. Buffa A, Dölz J, Kurz S, Schöps S, Vázquez R, Wolf F (2020) Multipatch approximation of the de Rham sequence and its traces in isogeometric analysis. Numerische Mathematik 144:201–236
6. Dölz J, Kurz S, Schöps S, Wolf F (2019) Isogeometric boundary elements in electromagnetism: rigorous analysis, fast methods, and examples. SIAM J Sci Comput 41(5):B983–B1010
7. Beirão da Veiga L, Buffa A, Sangalli G, Vázquez R (2014) Mathematical analysis of variational isogeometric methods. Acta Numer 23:157–287
8. Buffa A, Vázquez RH, Sangalli G, da Veiga LB (2015) Approximation estimates for isogeometric spaces in multipatch geometries. Numer Methods Partial Diff Equ 31(2):422–438
9. Harbrecht H, Randrianarivony M (2010) From computer aided design to wavelet BEM. Comput Vis Sci 13:69–82
10. Buffa A, Rivas J, Sangalli G, Vázquez R (2011) Isogeometric discrete differential forms in three dimensions. SIAM J Numer Anal 49(2):818–844

11. Peterson AF, Aberegg KR (1995) Parametric mapping of vector basis functions for surface integral equation formulations. Appl Comput Electromag Soc J 10:107–115
12. Bossavit A (1998) Computational electromagnetism: variational formulations, complementarity, edge elements. Academic, San Diego
13. Kurz S, Rain O, Rjasanow S (2002) The adaptive cross-approximation technique for the 3D boundary-element method. IEEE Trans Magn 38(2):421–424
14. Peterson AF (2006) Mapped vector basis functions for electromagnetic integral equations. Synth Lect Comput Electromagn 1(1):1–124
15. Monk P (2003) Finite element methods for Maxwell's equations. Oxford University Press, Oxford
16. Payne L, Weinberger H (1960) An optimal Poincaré inequality for convex domains. Arch Rat Mech Anal 5:286–292
17. Roustant O, Barthe F, Iooss B (2017) Poincaré inequalities on intervals - application to sensitivity analysis. Electr J Stat 11(2):3081–3119
18. Brenner SC, Scott LR (2008) The mathematical theory of finite element methods, vol 15, 3rd edn. Texts in applied mathematics. Springer, New York
19. Dölz J, Harbrecht H, Kurz S, Schöps S, Wolf F (2018) A fast isogeometric BEM for the three dimensional Laplace- and Helmholtz problems. Comput Methods Appl Mech Eng 330(Supplement C):83–101
20. Adams RA (1978) Sobolev spaces. Pure and applied mathematics. Academic, New York
21. Sauter SA, Schwab C (2011) Boundary element methods. Springer, Berlin
22. Buffa A, Christiansen S (2003) The electric field integral equation on Lipschitz screens: definitions and numerical approximation. Numerische Mathematik 94(2):229–267
23. Buffa A, Hiptmair R (2003) Galerkin boundary element methods for electromagnetic scattering. In: Ainsworth M, Davies P, Duncan D, Rynne B, Martin P (eds) Topics in computational wave propagation. Springer, Berlin, pp 83–124
24. Girault V, Raviart P-A (1986) Finite element methods for Navier-Stokes equations: theory and algorithms, 1st edn. Springer, Berlin
25. McLean W (2000) Strongly elliptic systems and boundary integral equations. Cambridge University Press, Cambridge
26. Boffi D, Brezzi F, Fortin M (2013) Mixed finite element methods and applications. Springer, Berlin
27. Costabel M, Dauge M, Demkowicz L (2008) Polynomial extension operators for H1, H(curl) and H(div)-spaces on a cube. Math Comput 1967–1999
28. Quarteroni A, Valli A (1999) Domain decomposition methods for partial differential equations, Numerical mathematics and scientific computation. Oxford University Press, Oxford
29. Xu J, Zikatanov L (2002) Some observations on Babuška and Brezzi theories. Numerische Mathematik 94(1):195–202
30. Bespalov A, Heuer N, Hiptmair R (2010) Convergence of the natural hp-BEM for the electric field integral equation on polyhedral surfaces. SIAM J Numer Anal 48:1518–1529
31. Hiptmair R (2002) Finite elements in computational electromagnetism. Acta Numer 11:237–339
32. Zaglmayr S (2006) High order finite element methods for electromagnetic field computation. Dissertation. Johannes Kepler Universität, Linz, Austria
33. Babuška I (1969) Error-bounds for finite element method. J Numer Math 16(4):322–333
34. Jackson JD (1998) Classical electrodynamics, 3rd edn. Wiley, New York

Chapter 4
Algorithmic Considerations for Matrix Assembly

In this chapter, we will review the algorithmic approach to the matrix assembly of the linear system induced by the isogeometric discretisation of the EFIE. Some of the material in this chapter was previously published in [1], and is reprinted with permission.[1]

Since, in general, the matrices will be densely populated due to the non-local kernels of the integral operators, a compression scheme needs to be utilised for large computations.

While adaptive cross approximation (ACA) techniques tailored to the isogeometric setting exist and yield good results, see e.g. [2], we must briefly comment on the classical black-box ACA, which has been used, for example, in [3]. Being a simple and purely algebraic algorithm, the ACA is very popular in the engineering community, cf. [4]. However, due to the overlapping supports of higher-order B-spline discretisations, it will inevitably lead to redundant kernel and geometry evaluations.

As an example, compare the lowest-order case in Fig. 4.1a to the higher-order case depicted in Fig. 4.1b. If one computes the interaction between one function φ_1 to two different functions φ_2, φ_3, evaluations of kernel and geometry on multiple elements will coincide in the higher-order case. These values would need to be either recomputed or stored, where one option is permissive because of its computational complexity and the other for its storage requirement.

Since in the isogeometric setting the support will scale with p, such a scheme will become increasingly inefficient for higher-order.

In contrast, the scheme we are going to present in this chapter is not based on the ACA. It is a non-trival variant of a compression scheme based on the interpolation of the kernel as introduced by [5] and adopted to scalar IGA-BEM in [6]. A particular

[1]Copyright ©2019 Society for Industrial and Applied Mathematics. Reprinted with permission. All rights reserved.

© The Author(s), under exclusive license to Springer Nature Switzerland AG 2021
F. Wolf, *Analysis and Implementation of Isogeometric Boundary Elements for Electromagnetism*, Springer Theses,
https://doi.org/10.1007/978-3-030-61939-8_4

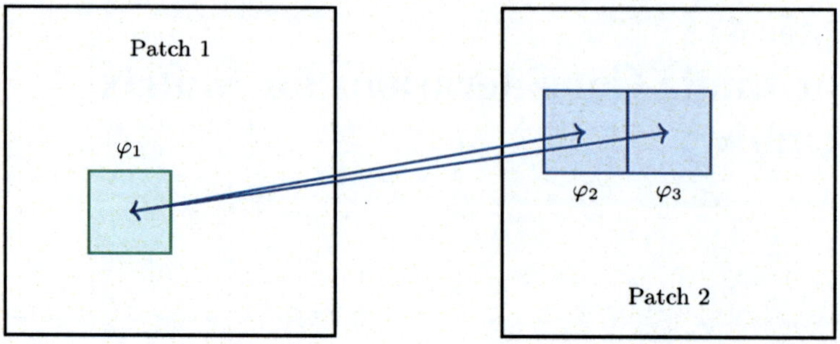

(a) Lower-order functions are local to each element. Due to the close element-to-basis-function relation evaluations of interactions may be computed between basis functions.

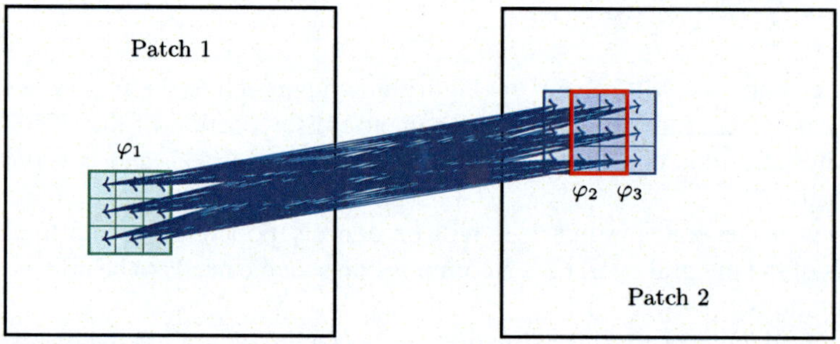

(b) Higher-order spline functions span their support over multiple elements. In this example consider scalar functions with $p = 3$ and thus a support of 3×3 elements. Due to higher choice of p, more quadrature points are required as well. The overlap contains six elements.

Fig. 4.1 A basis function oriented approach yields either a large overhead for bookkeeping or requires many redundant kernel and geometry evaluations

feature of this approach is that its implementation integrates effortlessly into the $\mathcal{H}^2$-matrix framework [7] and avoids redundant computations due to its element-based philosophy. This will be realised through what we call a *superspace approach*.

This chapter is structured as follows. First, in Sect. 4.1, we will review the representation of the Maxwell single layer operator w.r.t. the reference domain $\square$. These representations are used to assembly a local element matrix, which can then be translated into the system matrix by the superspace approach, which will be introduced in Sect. 4.2. Afterwards, in Sect. 4.3, we will introduce an interpolation-based multipole scheme which provably reduces the storage requirements and complexity of the matrix-vector multiplication of the local element matrix.

4.1 Integration on the Reference Domain

Before we introduce the compression scheme, we first have to pull the matrix representation (3.52) back to the reference domain. For this, we will first introduce some essential concepts.

We assume that the B-spline space on $\mathbb{S}^1_{\mathbf{p},\boldsymbol{\Xi}_m}(\Gamma)$ is generated by the tuple $\boldsymbol{\Xi}_{m,j} = (\Xi_m, \Xi_m)$ on each patch Γ_j, where Ξ_m is an equidistant knot vector with $2^m, m \geq 0$, elements.

This corresponds to m steps of uniform refinement in terms of the reference domain and generates a nested sequence of meshes. Then, for each level of refinement m, the mesh consists of 4^m elements per patch. The key point of this refinement strategy is that it induces a quadtree structure on the geometry, cf. Fig. 4.2, which the compression scheme will be based around. Each element within the nested sequence of meshes will be denoted by $\Gamma_{i,\tilde{m},k}$, i.e., referred to by a tuple $\boldsymbol{\lambda} := (i, \tilde{m}, k)$. In this, i denotes the corresponding parametric mapping, $\tilde{m} < m$ showcases the level of refinement of the element and k denotes the index of the element in hierarchical order. For notational purposes, we will define $|\boldsymbol{\lambda}| := \tilde{m}$ and also introduce the diffeomorphisms $\boldsymbol{F}_{\boldsymbol{\lambda}} \colon \square \to \Gamma_{\boldsymbol{\lambda}}$ which can easily be defined by combining $\boldsymbol{F}_i$ with a suitable affine transformation. For an efficient compression, each instance of $\Gamma_{\boldsymbol{\lambda}}$ is also considered as a *cluster*, in the sense that $\Gamma_{\boldsymbol{\lambda}}$ will be considered as the set of tree leaves appended to the subtree with root $\Gamma_{\boldsymbol{\lambda}}$. Naïvely said, $\Gamma_{\boldsymbol{\lambda}}$ can be visualised as "a square region on the geometry", compare Fig. 4.2a. The hierarchically ordered collection of all $\Gamma_{\boldsymbol{\lambda}}$ will be called *cluster tree* and denoted by $\mathcal{T}$.

For each pair of clusters in $\mathcal{T}$, the fundamental solution u^*_κ can be localised to a *localised fundamental solution*

$$u^*_{\kappa,\boldsymbol{\lambda},\boldsymbol{\lambda}'} \colon \square \times \square \to \mathbb{C}, \qquad u^*_{\kappa,\boldsymbol{\lambda},\boldsymbol{\lambda}'}(\boldsymbol{s}, \boldsymbol{t}) = u^*_\kappa\big(\boldsymbol{F}_{\boldsymbol{\lambda}}(\boldsymbol{s}) - \boldsymbol{F}_{\boldsymbol{\lambda}'}(\boldsymbol{t})\big) \tag{4.1}$$

which reparameterises the fundamental solution to $\square \times \square$, and localises the geometry mappings $\boldsymbol{F}_j$ to cluster-oriented mappings $\boldsymbol{F}_{\boldsymbol{\lambda}}$ through scaling and translation. This reduces the dimension (in terms of input variables) of the fundamental solution artificially to the dimension of the boundary.

On each element $\Gamma_{\boldsymbol{\lambda}}$, ansatz functions $\varphi_{\boldsymbol{\lambda}}$ are defined by lifting suitable shape functions $\widehat{\varphi}$ on $\square$ to the surface by the suitable localised pull-back.

To obtain suitable shape functions of polynomial degree p on $\square$, we introduce the knot vector Ξ^*_m, generated from Ξ_m by increasing the multiplicity of each knot to $p+1$. We then define the spaces $\mathbb{S}^*_{p,m}(\square)$, to be the discontinuous spaces generated by $\boldsymbol{p} = (p, p)$ and $\boldsymbol{\Xi}^*_m = (\Xi^*_m, \Xi^*_m)$, i.e., the space

$$\mathbb{S}^*_{p,m}(\square) := S_p(\Xi^*_m) \otimes S_p(\Xi^*_m). \tag{4.2}$$

By definition of the B-spline basis $\mathbb{S}^*_{p,m}(\square)$ contains exactly the tensor product Bernstein polynomials of degree p on each element, i.e., is locally of dimension $(p+1)^2$. The refinement yields 4^m elements per patch, and thus, after mapping $\mathbb{S}^*_{p,m}(\square)$ to

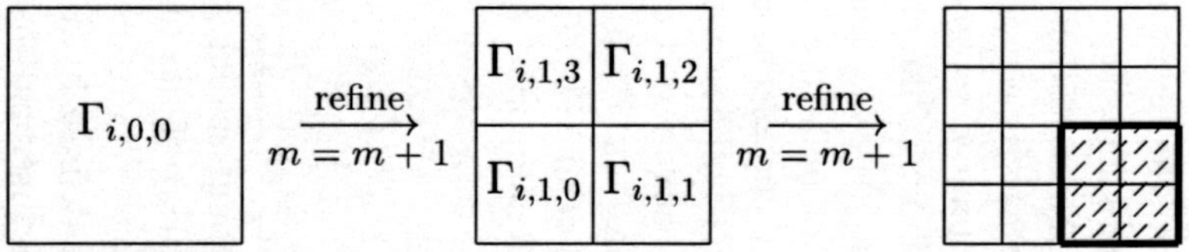

(a) Refinement of the patch induced by the i-th mapping. Bold region corresponds to cluster $\Gamma_{\boldsymbol{\lambda}}$ with $\boldsymbol{\lambda} = (i, 1, 1)$.

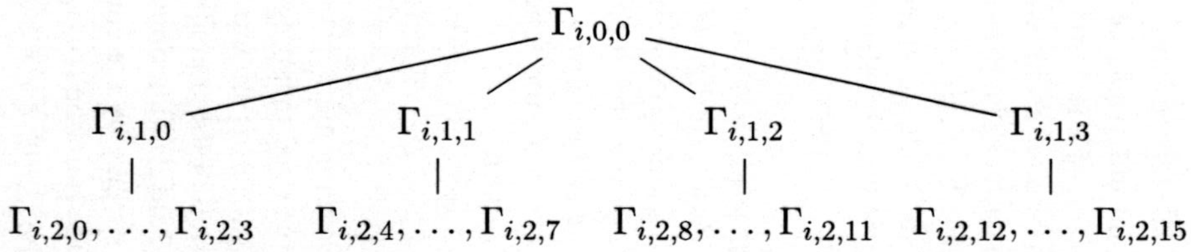

(b) Quadtree structure induced by refinement of the i-th mapping.

Fig. 4.2 Refinement, labelling, and quadtree structure. Reprinted from [6], Copyright ©2018, with permission from Elsevier each patch Γ_j, this yields a global discrete discontinuous function space $\mathbb{S}^*_{p,m}(\Gamma)$ of dimension $4^m N(p+1)^2$, where N refers to the number of patches. According to [8], for two basis functions $\boldsymbol{\varphi}_i$ and $\boldsymbol{\varphi}_j$ of $\mathbf{S}^*_{p,m}(\Gamma)$ supported on $\Gamma_{\boldsymbol{\lambda}(i)}$ and $\Gamma_{\boldsymbol{\lambda}(j)}$, together with their pulled-back counterparts $\widehat{\boldsymbol{\varphi}}_i = \boldsymbol{\varphi}_i \circ \boldsymbol{F}_{\boldsymbol{\lambda}_i}$, $\widehat{\boldsymbol{\varphi}}_j = \boldsymbol{\varphi}_j \circ \boldsymbol{F}_{\boldsymbol{\lambda}_j}$, the first integral of the Maxwell single layer operator is given by

$$\begin{aligned}\int_\Gamma \int_\Gamma & u^*_\kappa(\boldsymbol{x}-\boldsymbol{y})\boldsymbol{\varphi}_i(\boldsymbol{x}) \cdot \boldsymbol{\varphi}_j(\boldsymbol{y}) \,\mathrm{d}\Gamma_{\boldsymbol{y}} \,\mathrm{d}\Gamma_{\boldsymbol{x}} \\ &= \int_\square \int_\square u^*_{\kappa,\boldsymbol{\lambda}(i),\boldsymbol{\lambda}(j)}(\boldsymbol{s},\boldsymbol{t}) \langle \boldsymbol{dF}_{\boldsymbol{\lambda}(j)}(\boldsymbol{s})\widehat{\boldsymbol{\varphi}}_j(\boldsymbol{s}), \boldsymbol{dF}_{\boldsymbol{\lambda}(i)}(\boldsymbol{t})\widehat{\boldsymbol{\varphi}}_i(\boldsymbol{t})\rangle \,\mathrm{d}\boldsymbol{t}\,\mathrm{d}\boldsymbol{s},\end{aligned} \tag{4.3}$$

and the second by

$$\begin{aligned}\int_\Gamma \int_\Gamma & u^*_\kappa(\boldsymbol{x}-\boldsymbol{y}) \operatorname{div}_\Gamma \boldsymbol{\varphi}_j(\boldsymbol{x}) \operatorname{div}_\Gamma \boldsymbol{\varphi}_i(\boldsymbol{y}) \,\mathrm{d}\Gamma_{\boldsymbol{y}} \,\mathrm{d}\Gamma_{\boldsymbol{x}} \\ &= \int_\square \int_\square u^*_{\kappa,\boldsymbol{\lambda}(i),\boldsymbol{\lambda}(j)}(\boldsymbol{s},\boldsymbol{t}) \operatorname{div} \widehat{\boldsymbol{\varphi}}_j(\boldsymbol{s}) \operatorname{div} \widehat{\boldsymbol{\varphi}}_i(\boldsymbol{t}) \,\mathrm{d}\boldsymbol{t}\,\mathrm{d}\boldsymbol{s}.\end{aligned} \tag{4.4}$$

Let, for the remainder of this chapter, ∂_1 and ∂_2 denote the derivation w.r.t. the first or second argument.

Assuming that a finite dimensional basis of $\mathbf{S}^*_{p,m}(\Gamma)$ is given in terms of scalar functions, i.e.,

$$\left\{ \begin{bmatrix} \varphi_i \\ 0 \end{bmatrix}, \begin{bmatrix} 0 \\ \varphi_j \end{bmatrix} : \varphi_i, \varphi_j \text{ basis functions of } \mathbb{S}^*_{p,m}(\Gamma) \right\},$$

with their pulled-back counterparts $\widehat{\varphi}_i, \widehat{\varphi}_j$, the matrix $\boldsymbol{V}^*_{\kappa,h}$ can be further decomposed into

$$\boldsymbol{V}^*_{\kappa,h} = \begin{bmatrix} \boldsymbol{V}^{(1,1)}_{\kappa,h} & \boldsymbol{V}^{(1,2)}_{\kappa,h} \\ \boldsymbol{V}^{(2,1)}_{\kappa,h} & \boldsymbol{V}^{(2,2)}_{\kappa,h} \end{bmatrix},$$

with

$$\begin{aligned}\left[\boldsymbol{V}^{(\alpha,\beta)}_{\kappa,h}\right]_{i,j} = \int_\square \int_\square u^*_{\kappa,\boldsymbol{\lambda}(i),\boldsymbol{\lambda}(j)}(\boldsymbol{s},\boldsymbol{t}) \Big(& \langle \partial_\alpha \boldsymbol{F}_{\boldsymbol{\lambda}(i)}(\boldsymbol{s}), \partial_\beta \boldsymbol{F}_{\boldsymbol{\lambda}(j)}(\boldsymbol{t})\rangle \widehat{\varphi}_j(\boldsymbol{s})\widehat{\varphi}_i(\boldsymbol{t}) \\ & - \frac{1}{\kappa^2} \partial_\alpha \widehat{\varphi}_j(\boldsymbol{s}) \partial_\beta \widehat{\varphi}_i(\boldsymbol{t}) \Big) \,\mathrm{d}\boldsymbol{t}\,\mathrm{d}\boldsymbol{s},\end{aligned} \tag{4.5}$$

for $\alpha, \beta = 1, 2$.

Remark 4.1 (*Symmetry of the Bilinear Form*) To obtain an efficient implementation, one may choose to simultaneously assemble the $\boldsymbol{V}^{(\alpha,\beta)}_{\kappa,h}$ and exploit the symmetry $\boldsymbol{V}^{(2,1)}_{\kappa,h} = \big(\boldsymbol{V}^{(1,2)}_{\kappa,h}\big)^\intercal$ and the intrinsic symmetry of the blocks $\boldsymbol{V}^{(1,1)}_{\kappa,h}$ and $\boldsymbol{V}^{(2,2)}_{\kappa,h}$.

Remark 4.2 (*Element-Wise Quadrature*) Choosing to compute element-element interactions instead of function-function interaction avoids redundant kernel and geometry evaluations by design. This comes naturally with the presented superspace approach. On each element, we employ classical boundary element schemes for numerical integration of singular integrals, often referred to as *Sauter–Schwab quadrature rules*, see [9] or [10]. These can be applied without modifications.

4.2 The Superspace Approach

Since B-splines are piecewise polynomials the relation

$$\mathbb{S}^1_{\boldsymbol{p},\boldsymbol{\Xi}_m}(\Gamma) \subseteq \mathbb{S}^*_{\boldsymbol{p},m}(\Gamma) := \mathbb{S}^*_{p,m}(\Gamma) \times \mathbb{S}^*_{p,m}(\Gamma)$$

with $\boldsymbol{p} = (p, p)$ and $\boldsymbol{\Xi}_m = (\Xi_m, \Xi_m)$ is evident. We can therefore represent each basis function of $\mathbb{S}^1_{\boldsymbol{p},\boldsymbol{\Xi}_m}(\Gamma)$ by a linear combination of basis functions of $\mathbb{S}^*_{\boldsymbol{p},m}(\Gamma)$.

Now, instead of assembling the system of linear equations (3.51) with respect to $\mathbb{S}^1_{\boldsymbol{p},\boldsymbol{\Xi}_m}(\Gamma)$, one may assemble it with respect to $\mathbb{S}^*_{\boldsymbol{p},m}(\Gamma)$ to obtain a local element matrix $\boldsymbol{V}^*_{\kappa,h}$ and a vector $\boldsymbol{f}^*_h$. A linear system of equations equivalent to (3.51) is then given by

$$\underbrace{\boldsymbol{T}^\intercal \boldsymbol{V}^*_{\kappa,h} \boldsymbol{T}}_{=\boldsymbol{V}_{\kappa,h}}\, \boldsymbol{w} = -\boldsymbol{T}^\intercal \boldsymbol{f}^*_h \tag{4.6}$$

for a suitable transformation matrix $\boldsymbol{T}$.

Remark 4.3 (*Computation of* $\boldsymbol{T}$) Set $|\mathbb{S}^*| = \dim(\mathbb{S}^*_{\boldsymbol{p},m}(\Gamma))$. The coefficients encoded in T are exactly the coefficients c_i required to represent

$$\begin{aligned} a(\boldsymbol{b}_j, \boldsymbol{b}_i) &= a\left(\sum_{0\le j<|\mathbb{S}^*|} c_j \tilde{\boldsymbol{b}}_j, \sum_{0\le i<|\mathbb{S}^*|} c_i \tilde{\boldsymbol{b}}_i\right) \\ &= \sum_{0\le j<|\mathbb{S}^*|} c_j a\left(\tilde{\boldsymbol{b}}_j, \sum_{0\le j<|\mathbb{S}^*|} c_i \tilde{\boldsymbol{b}}_i\right) \\ &= \sum_{0\le j<|\mathbb{S}^*|} c_i \sum_{0\le i<|\mathbb{S}^*|} c_j a\left(\tilde{\boldsymbol{b}}_j, \tilde{\boldsymbol{b}}_i\right), \end{aligned}$$

for all $\boldsymbol{b}_j, \boldsymbol{b}_i \in \mathbb{S}^1_{\boldsymbol{p},\boldsymbol{\Xi}}(\Gamma)$ and $\tilde{\boldsymbol{b}}_j, \tilde{\boldsymbol{b}}_i \in \mathbb{S}^*_{\boldsymbol{p},m}(\Gamma)$, and any bilinear form a. Note that most of the coefficients will be zero. This is exactly the relation utilised in (4.6), where the double sum is written as a basis transformation.

In the implementation, the coefficients are computed by solving multiple interpolation problems a priori in a black-box fashion: Let $\boldsymbol{b}_i \in \mathbb{S}^1_{\boldsymbol{p},\boldsymbol{\Xi}}(\Gamma)$ be fixed and choose $(p+1)^2$ suitable interpolation points $\boldsymbol{x}_i \in \Gamma$ on each element. Then, for $\boldsymbol{b} = (\boldsymbol{b}_i(\boldsymbol{x}_0), \ldots, \boldsymbol{b}_i(\boldsymbol{x}_{k-1}))^\intercal$, the ith column of $\boldsymbol{T}$ is given by a vector $\boldsymbol{c}$ which is the solution to

$$Ac = b,$$

with $A_{i,j} = \tilde{b}_i(x_j)$.

Note that the structure of this interpolation problem is local to each element and can be pulled back to the reference element. Thus, only local problems of size $(p+1) \times (p+1)$ must be solved. For this, the inverse of this small interpolation matrix must be computed only once.

Remark 4.4 (*Compression of* $V^*_{\kappa,h}$) Since the dimension of $\mathbb{S}^*_{p,m}(\Gamma)$ is larger than the dimension of $\mathbb{S}^1_{p,\Xi_m}(\Gamma)$, the matrix $V^*_{\kappa,h}$ is larger than the matrix $V_{\kappa,h}$. However, it has been shown in [11] for the case of classical higher-order Raviart–Thomas elements that the superspace approach can achieve better compression rates and, thus, better computation times.

Indeed, the local support of $\mathbb{S}^*_{p,m}(\Gamma)$ will benefit the compression scheme immensely, as will be discussed later on.

4.3 Interpolation-Based Multipole Method

Due to the non-locality of the fundamental solution u^*_κ, the local element matrix $V^*_{\kappa,h}$ is densely populated. Its storage and assembly cost are thus prohibitively expensive for higher-dimensional ansatz and test spaces, and an efficient fast method for the reduction of storage requirements and the cost of matrix-vector products is needed.

A particular feature of this approach is that its simple implementation integrates effortlessly into the $\mathcal{H}^2$-matrix framework [7], which is a more efficient variant of the commonly used $\mathcal{H}$-matrix variant [12].

We follow the approach of [5, 6, 13] to compress the matrices $V^{(\alpha,\beta)}_{\kappa,h}$, $\alpha, \beta = 1, 2$, in terms of a specialised fast multipole method, which yields a representation of these matrices in terms of $\mathcal{H}^2$-matrices, see also [7]. However, the approach is only applicable to matrices of the kind

$$\begin{aligned}[A]_{i,j} &= \int_\Gamma \int_\Gamma u^*_\kappa(x-y)\varphi_j(x)\varphi_i(y)\,\mathrm{d}\sigma_y\,\mathrm{d}\sigma_x \\ &= \int_\square \int_\square u^*_{\kappa,\lambda(i),\lambda(j)}(s,t)\widehat{\varphi}_j(s)\widehat{\varphi}_i(t)\,\mathrm{d}t\,\mathrm{d}s,\end{aligned}$$

which does not readily fit the format of the matrices from (4.5) due to the derivatives of the geometry mappings contained in the basis functions and the involved surface divergences. In the following, we will, therefore, adapt the construction to the setting of the electric single layer operator.

For constructing the $\mathcal{H}^2$-matrix representation, consider the level-wise Cartesian product $\mathcal{T} \boxtimes \mathcal{T} := \{\Gamma_\lambda \times \Gamma_{\lambda'} \colon \Gamma_\lambda, \Gamma_{\lambda'} \in \mathcal{T}, |\lambda| = |\lambda'|\}$ of the cluster tree $\mathcal{T}$. Compressible matrix blocks are then identified by the following *admissibility condition*.

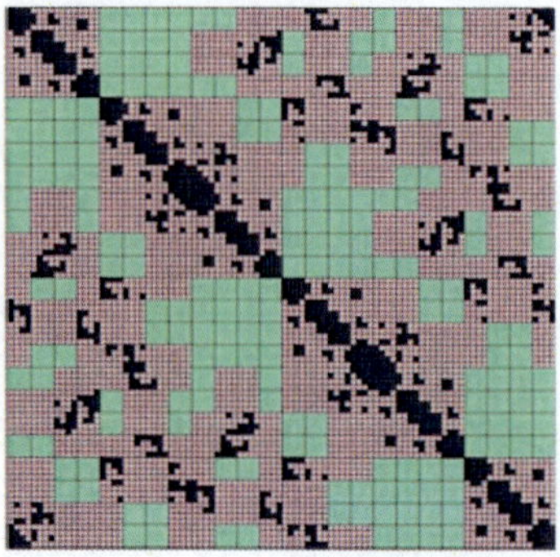

Fig. 4.3 Illustration of the $\mathcal{H}^2$-matrix partitioning. All but the darkest blocks are contained in the farfield and will be compressed. Copyright ©2019 Society for Industrial and Applied Mathematics. Reprinted with permission. All rights reserved

Definition 4.5 (*Admissibility Condition,* [13, Definition 5.1]) The clusters $\Gamma_{\boldsymbol{\lambda}}$ and $\Gamma_{\boldsymbol{\lambda}'}$ with $|\boldsymbol{\lambda}| = |\boldsymbol{\lambda}'|$ are called *admissible* if

$$\max\left\{\operatorname{diam}(\Gamma_{\boldsymbol{\lambda}}), \operatorname{diam}(\Gamma_{\boldsymbol{\lambda}'})\right\} \leq \eta \operatorname{dist}(\Gamma_{\boldsymbol{\lambda}}, \Gamma_{\boldsymbol{\lambda}'}) \tag{4.7}$$

holds for a fixed $\eta \in (0, 1)$. Let $\operatorname{pred}(\boldsymbol{\lambda})$ denote the predecessor of a tuple $\boldsymbol{\lambda}$ w.r.t. underlying tree structure of the refinement process. The largest collection of admissible blocks $\Gamma_{\boldsymbol{\lambda}} \times \Gamma_{\boldsymbol{\lambda}'} \in \mathcal{T} \boxtimes \mathcal{T}$ such that $\Gamma_{\operatorname{pred}(\boldsymbol{\lambda})} \times \Gamma_{\operatorname{pred}(\boldsymbol{\lambda}')}$ is not admissible will be called *far-field* $\mathcal{F} \subset \mathcal{T} \boxtimes \mathcal{T}$. The remaining non-admissible blocks will be called *near-field* $\mathcal{N} \subset \mathcal{T} \boxtimes \mathcal{T}$.

The far-field yields the compressible matrix blocks, while the near-field is assembled densely, see Fig. 4.3 for an illustration. The *block-cluster tree* $\mathcal{B} := \mathcal{F} \cup \mathcal{N}$ can be constructed as explained in [6]. We remark that for all block-clusters $\Gamma_{\boldsymbol{\lambda}} \times \Gamma_{\boldsymbol{\lambda}'} \in \mathcal{B}$, it holds $|\boldsymbol{\lambda}| = |\boldsymbol{\lambda}'|$ and refer to [5, 13] for an in-depth discussion about the special properties of the block-cluster tree in the isogeometric setting.

For a given polynomial degree $q \in \mathbb{N}$, let $\{x_0, x_1, \ldots, x_q\} \subset [0, 1]$ denote $q+1$ interpolation points. Furthermore, let $L_r(s)$ for $r = 0, \ldots, q$ be the Lagrangian basis polynomials with respect to these interpolation points. By a tensor product construction, one obtains the interpolation points $\boldsymbol{x_r} := (x_{r_1}, x_{r_2})$ and the corresponding tensor product basis polynomials

$$L_{\boldsymbol{r}}(\boldsymbol{s}) := L_{r_1}(s_1) \cdot L_{r_2}(s_2) \quad \text{for } r_1, r_2 = 0, \ldots, q,$$

in all admissible blocks $\Gamma_{\boldsymbol{\lambda}} \times \Gamma_{\boldsymbol{\lambda}'} \in \mathcal{F}$. This gives rise to the approximation

$$u^*_{\kappa,\boldsymbol{\lambda},\boldsymbol{\lambda}'}(\boldsymbol{s}, \boldsymbol{t}) \approx \sum_{\substack{\|\boldsymbol{r}\|_\infty \leq q,\\ \|\boldsymbol{r}'\|_\infty \leq q}} u^*_{\kappa,\boldsymbol{\lambda},\boldsymbol{\lambda}'}(\boldsymbol{x_r}, \boldsymbol{x_{r'}}) L_{\boldsymbol{r}}(\boldsymbol{s}) L_{\boldsymbol{r}'}(\boldsymbol{t}).$$

Remark 4.6 (*Complexity of the Multipole Approximation*) We remark that the approach presented here interpolates the localised kernel (4.1) via polynomials on the reference domain $\square$ of the isogeometric mappings. Specifically, this amounts to an interpolation on a two-dimensional manifold, rather than the original kernel in space, i.e, interpolation in three-dimensional space. The latter was first introduced in [14, 15]. The intrinsically two-dimensional approach has complexity of $q^4 = q^2 \cdot q^2$ instead of a complexity of $q^6 = q^3 \cdot q^3$ for the volume case, in terms of the interpolation degree q of the multipole scheme.

We remind ourselves that $S_p(\Xi_m^*)$ denotes the spline space based on Ξ_m for maximal knot repetition, i.e, of Bernstein polynomial on each element of Ξ_m, compare Eq. (4.2). Including the geometry information into the kernel evaluation yields

$$\boldsymbol{V}_{\kappa,h}^{(\alpha,\beta)}\Big|_{\boldsymbol{\lambda},\boldsymbol{\lambda}'} = \boldsymbol{V}_{\kappa,h,1}^{(\alpha,\beta)}\Big|_{\boldsymbol{\lambda},\boldsymbol{\lambda}'} + \boldsymbol{V}_{\kappa,h,2}^{(\alpha,\beta)}\Big|_{\boldsymbol{\lambda},\boldsymbol{\lambda}'} \tag{4.8}$$

with

$$\begin{aligned}
&\Big[\boldsymbol{V}_{\kappa,h,1}^{(\alpha,\beta)}\Big|_{\boldsymbol{\lambda},\boldsymbol{\lambda}'}\Big]_{\ell,\ell'} \\
&= \int_\square \int_\square u^*_{\kappa,\boldsymbol{\lambda},\boldsymbol{\lambda}}(\boldsymbol{s},\boldsymbol{t}) \langle \partial_\alpha \boldsymbol{F}_{\boldsymbol{\lambda}}(\boldsymbol{s}), \partial_\beta \boldsymbol{F}_{\boldsymbol{\lambda}}(\boldsymbol{t})\rangle \widehat{\varphi}_{\ell'}(\boldsymbol{s}) \widehat{\varphi}_{\ell}(\boldsymbol{t}) \,\mathrm{d}\boldsymbol{t}\,\mathrm{d}\boldsymbol{s} \\
&\approx \sum_{\substack{\|\boldsymbol{r}\|_\infty \le q,\\ \|\boldsymbol{r}'\|_\infty \le q}} u^*_{\kappa,\boldsymbol{\lambda},\boldsymbol{\lambda}'}(\boldsymbol{x}_{\boldsymbol{r}}, \boldsymbol{x}_{\boldsymbol{r}'}) \langle \partial_\alpha \boldsymbol{F}_{\boldsymbol{\lambda}}(\boldsymbol{x}_{\boldsymbol{r}}), \partial_\beta \boldsymbol{F}_{\boldsymbol{\lambda}}(\boldsymbol{x}_{\boldsymbol{r}'})\rangle \\
&\qquad \cdot \int_\square L_{\boldsymbol{r}}(\boldsymbol{s}) \widehat{\varphi}_{\ell'}(\boldsymbol{s}) \,\mathrm{d}\boldsymbol{s} \int_\square L_{\boldsymbol{r}'}(\boldsymbol{t}) \widehat{\varphi}_{\ell}(\boldsymbol{t}) \,\mathrm{d}\boldsymbol{t}
\end{aligned}$$

for two basis functions $\widehat{\varphi}_\ell, \widehat{\varphi}_{\ell'} \in \mathbb{S}^*_{p,m-|\boldsymbol{\lambda}|}(\square)$. This means we have the representation

$$\Big[\boldsymbol{V}_{\kappa,h,1}^{(\alpha,\beta)}\Big|_{\boldsymbol{\lambda},\boldsymbol{\lambda}'}\Big]_{\ell,\ell'} = \big[\boldsymbol{M}^\square_{|\boldsymbol{\lambda}|} \boldsymbol{K}^{(\alpha,\beta)}_{\boldsymbol{\lambda},\boldsymbol{\lambda}',1} (\boldsymbol{M}^\square_{|\boldsymbol{\lambda}'|})^\intercal\big]_{\ell,\ell'},$$

where

$$\Big[\boldsymbol{K}^{(\alpha,\beta)}_{\boldsymbol{\lambda},\boldsymbol{\lambda}',1}\Big]_{\boldsymbol{r},\boldsymbol{r}'} = u^*_{\kappa,\boldsymbol{\lambda},\boldsymbol{\lambda}'}(\boldsymbol{x}_{\boldsymbol{r}}, \boldsymbol{x}_{\boldsymbol{r}'}) \langle \partial_\alpha \boldsymbol{F}_{\boldsymbol{\lambda}}(\boldsymbol{x}_{\boldsymbol{r}}), \partial_\beta \boldsymbol{F}_{\boldsymbol{\lambda}}(\boldsymbol{x}_{\boldsymbol{r}'})\rangle$$

and

$$\begin{aligned}
\big[\boldsymbol{M}_{|\boldsymbol{\lambda}|}\big]_{r,\ell} &= \int_0^1 L_r(s) \widehat{\phi}_\ell(s)\,\mathrm{d}s, \quad \widehat{\phi}_\ell \in S_p(\Xi^*_{m-|\boldsymbol{\lambda}|}), \\
\boldsymbol{M}^\square_{|\boldsymbol{\lambda}|} &= \boldsymbol{M}_{|\boldsymbol{\lambda}|} \otimes \boldsymbol{M}_{|\boldsymbol{\lambda}|}.
\end{aligned}$$

For the second term in (4.8) we obtain

$$\begin{aligned}
&\Big[\boldsymbol{V}_{\kappa,h,2}^{(\alpha,\beta)}\Big|_{\boldsymbol{\lambda},\boldsymbol{\lambda}'}\Big]_{\ell,\ell'} \\
&= -\frac{1}{\kappa^2} \int_\square \int_\square u^*_{\kappa,\boldsymbol{\lambda},\boldsymbol{\lambda}}(\boldsymbol{s},\boldsymbol{t}) \partial_\alpha \widehat{\varphi}_{\ell'}(\boldsymbol{s}) \partial_\beta \widehat{\varphi}_{\ell}(\boldsymbol{t}) \,\mathrm{d}\boldsymbol{t}\,\mathrm{d}\boldsymbol{s} \\
&\approx \sum_{\substack{\|\boldsymbol{r}\|_\infty \le q,\\ \|\boldsymbol{r}'\|_\infty \le q}} -\frac{1}{\kappa^2} u^*_{\kappa,\boldsymbol{\lambda},\boldsymbol{\lambda}'}(\boldsymbol{x}_{\boldsymbol{r}}, \boldsymbol{x}_{\boldsymbol{r}'}) \int_\square L_{\boldsymbol{r}}(\boldsymbol{s}) \partial_\alpha \widehat{\varphi}_{\ell'}(\boldsymbol{s}) \,\mathrm{d}\boldsymbol{s} \int_\square L_{\boldsymbol{r}'}(\boldsymbol{t}) \partial_\beta \widehat{\varphi}_{\ell}(\boldsymbol{t}) \,\mathrm{d}\boldsymbol{t},
\end{aligned}$$

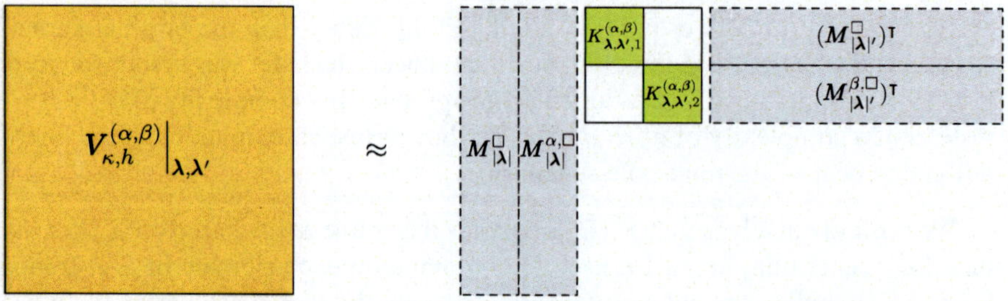

Fig. 4.4 Illustration of the compression of an admissible block $V_{\kappa,h}^{(\alpha,\beta)}\big|_{\lambda,\lambda'}$. Copyright ©2019 Society for Industrial and Applied Mathematics. Reprinted with permission. All rights reserved

which amounts to the representation

$$\Big[V_{\kappa,h,2}^{(\alpha,\beta)}\Big|_{\lambda,\lambda'}\Big]_{\ell,\ell'} = \big[M_{|\lambda|}^{\alpha,\square} K_{\lambda,\lambda',2}^{(\alpha,\beta)} (M_{|\lambda'|}^{\beta,\square})^{\intercal}\big]_{\ell,\ell'},$$

with

$$\begin{aligned}
\big[M_{|\lambda|}^{\partial}\big]_{r,\ell} &= \int_0^1 L_r(s)\partial\widehat{\phi}_\ell(s)\,\mathrm{d}s, \quad \widehat{\phi}_\ell \in S_p(\Xi_{m-|\lambda|}^*), \\
M_{|\lambda|}^{1,\square} &= M_{|\lambda|}^{\partial} \otimes M_{|\lambda|}, \\
M_{|\lambda|}^{2,\square} &= M_{|\lambda|} \otimes M_{|\lambda|}^{\partial},
\end{aligned}$$

and

$$\Big[K_{\lambda,\lambda',2}^{(\alpha,\beta)}\Big]_{r,r'} = -\frac{1}{\kappa^2} u_{\kappa,\lambda,\lambda'}^*(x_r, x_{r'}).$$

In view of (4.8), this yields the low-rank representation

$$V_{\kappa,h}^{(\alpha,\beta)}\Big|_{\lambda,\lambda'} \approx \Big[M_{|\lambda|}^{\square}\ M_{|\lambda|}^{\alpha,\square}\Big] \begin{bmatrix} K_{\lambda,\lambda',1}^{(\alpha,\beta)} & 0 \\ 0 & K_{\lambda,\lambda',2}^{(\alpha,\beta)} \end{bmatrix} \begin{bmatrix} (M_{|\lambda'|}^{\square})^{\intercal} \\ (M_{|\lambda'|}^{\beta,\square})^{\intercal} \end{bmatrix}, \tag{4.9}$$

for the matrices (4.5) in all admissible matrix blocks, see also Fig. 4.4 for an illustration.

We remark that this representation is within the same framework as it was used for the treatment of the hypersingular operator for the Laplace equation in [13]. Therefore, all considerations made in [13] apply. This immediately transfers the following result to our setting.

Theorem 4.7 (Algorithmic Complexity of the Compression, [13, Sect. 5.3]) *Let N denote the number of patches and m the level of refinement. The storage consumption of the compressed matrix has a complexity of $\mathcal{O}(N \cdot 4^m (pq)^2)$. Moreover, the matrix-vector multiplication has also a complexity of $\mathcal{O}(N \cdot 4^m (pq)^2)$, if a fast $\mathcal{H}^2$-implementation is used.*

It must be stressed that the introduced compression scheme has an intrinsic $\mathcal{H}^2$-structure. This renders it more efficient than the frequently used $\mathcal{H}$-matrix-based schemes. Its efficiency is based on the fact that, for each admissible block $\boldsymbol{\lambda} \times \boldsymbol{\lambda}'$, there are only q^4 evaluations of the geometry and the kernel function required to assemble the matrices $\boldsymbol{K}^{(\alpha,\beta)}_{\boldsymbol{\lambda},\boldsymbol{\lambda}',1}$ and $\boldsymbol{K}^{(\alpha,\beta)}_{\boldsymbol{\lambda},\boldsymbol{\lambda}',2}$. The other required matrices from (4.9) can be efficiently represented by recurrence relations from smaller matrices with tensor product structure such that assembly, storage and application do not affect the asymptotic behaviour, see [13].

An error analysis of this scheme has been conducted by my coauthors in [1] and yields the following.

Theorem 4.8 (Error of the Compression, [1, Theorem 4.7]) *The presented compression scheme maintains the existence and uniqueness of solutions of the numerical scheme. Moreover, there exists $q_0 > 0$ such that the optimal convergence rate of Theorem 3.48 is maintained if one chooses $q \sim (s + 5/2)m$ and $q \geq q_0$.*

Remark 4.9 (*Smoothness of the Geometry Mapping*) Note that both the analysis of Chap. 3 as well as the compression requires C^∞-mappings to the geometry. While NURBS mappings will, in general, only be C^k due to knot repetition, we employ Bézier extraction [16] as a preprocessing step of the geometry. This is the process of extracting locally C^∞-sub-patches of the geometry, which enables the geometry evaluation to be reduced to the evaluation of (rational) Bernstein polynomials. These sub-patches form the basis of the cluster tree for the proposed multipole method. By the superspace approach, smooth functions across multiple of these sub-patches can then be reconstructed through suitable coefficients within the matrix $\boldsymbol{T}$.

References

1. Dölz J, Kurz S, Schöps S, Wolf F (2019) Isogeometric boundary elements in electromagnetism: rigorous analysis, fast methods, and examples. SIAM J Sci Comput 41(5):B983–B1010
2. Marussig B, Zechner J, Beer G, Fries T-P (2015) Fast isogeometric boundary element method based on independent field approximation. Comput Meth Appl Mech Eng 284:458–488. Isogeometric Analysis Special Issue
3. Simpson R, Liu Z, Vázquez R, Evans J (2018) An isogeometric boundary element method for electromagnetic scattering with compatible B-spline discretizations. J Comput Phys 362:264–289
4. Kurz S, Rain O, Rjasanow S (2007) Fast boundary element methods in computational electromagnetism. In: Schanz M, Steinbach O (eds) Boundary element analysis: mathematical aspects and applications. Springer, Berlin, pp 249–279
5. Harbrecht H, Peters M (2013) Comparison of fast boundary element methods on parametric surfaces. Comput Methods Appl Mech Eng 261:39–55
6. Dölz J, Harbrecht H, Kurz S, Schöps S, Wolf F (2018) A fast isogeometric BEM for the three dimensional Laplace- and Helmholtz problems. Comput Methods Appl Mech Eng 330(Supplement C):83–101
7. Börm S (2010) Efficient numerical methods for non-local operators, vol 14. EMS tracts in mathematics. European Mathematical Society (EMS), Zürich

8. Peterson AF, Aberegg KR (1995) Parametric mapping of vector basis functions for surface integral equation formulations. Appl Comput Electromag Soc J 10:107–115
9. Sauter SA, Schwab C (2011) Boundary element methods. Springer, Berlin
10. Harbrecht H (2011) Wavelet Galerkin schemes for the boundary element method in three dimensions. Dissertation. Technische Universität Chemnitz
11. Rjasanow S, Weggler L (2017) Matrix valued adaptive cross approximation. Math Methods Appl Sci 40:2522–2531
12. Hackbusch W (2015) Hierarchical matrices: algorithms and analysis. Springer, Berlin
13. Dölz J, Harbrecht H, Peters M (2016) An interpolation-based fast multipole method for higher-order boundary elements on parametric surfaces. Int J Numer Methods Eng 108(13):1705–1728
14. Giebermann K (2001) Multilevel approximation of boundary integral operators. Computing 67(3):183–207
15. Hackbusch W, Börm S (2002) $\mathcal{H}^2$-matrix approximation of integral operators by interpolation. Appl Numer Math 43(1):129–143
16. Borden MJ, Scott MA, Evans JA, Hughes TJR (2011) Isogeometric finite element data structures based on Bezier extraction of NURBS. Int J Numer Methods Eng 87(1–5):15–47

Chapter 5
Numerical Examples: Electromagnetic Scattering

This chapter is devoted to the presentation of multiple numerical examples. These are designed to verify the behaviour predicted in Theorem 3.48 and Corollary 3.49, and to benchmark the method.

Our approach to this is as follows. We will first investigate an example with a known analytical solution of the surface current, i.e., the unknown of the EFIE, thus verifying Theorem 3.48. We will also investigate multiple examples with a known solution to the electric field to check for the behaviour predicted by Corollary 3.49.

Afterwards, in Sect. 5.4, we will compare the B-Spline basis $\mathbb{S}^1_{\boldsymbol{p},\boldsymbol{\Xi}}(\Gamma)$ to a corresponding quadrilateral Raviart–Thomas discretisation. Section 5.5 will then showcase the case where only p-refinement is applied, which corresponds to what is sometimes called *spectral element method*, cf. [1]. Finally, the compression rates of the multipole method will be reviewed in Sect. 5.6.

Note that a commodity of fast boundary element methods is that they all rely on iterative solvers and, thus, they are likely to struggle with high condition numbers caused by a large product of wavenumber to geometry diameter. Thus, the arising systems are solved via a complex GMRES unless otherwise stated, without the application of preconditioners, since a treatise of preconditioning remains an open research topic.

As mentioned before, the geometry evaluation incorporates Bézier extraction for efficient geometry evaluations. Matrix assembly, matrix-vector multiplication and potential evaluation are parallelised via OpenMP [2]. The implementation, called Bembel, is publicly available [3] and is reviewed in Appendix.

© The Author(s), under exclusive license to Springer Nature Switzerland AG 2021

F. Wolf, *Analysis and Implementation of Isogeometric Boundary Elements for Electromagnetism*, Springer Theses,
https://doi.org/10.1007/978-3-030-61939-8_5

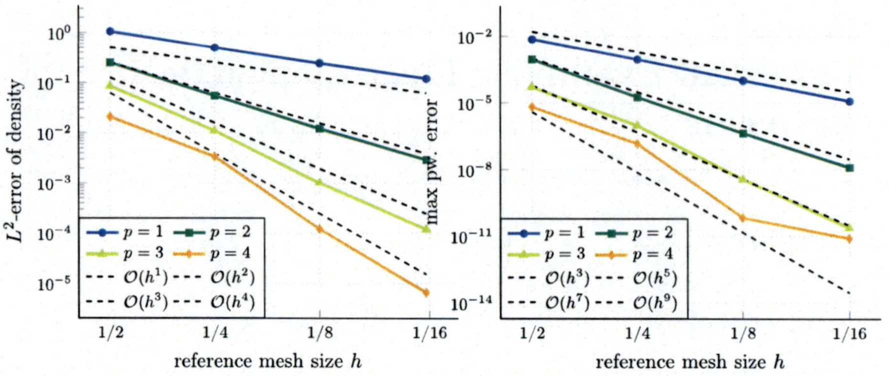

(a) $L^2(\Gamma)$-error of the density.

(b) Maximal pointwise error of electric field for $\mathbf{DP}_{(0,0.1,0.1)}$.

Fig. 5.1 Numerical examples on the unit sphere. Wavenumber $\kappa = 1$, parameters $q = 10$, and $\eta = 1.6$. The evaluation points are a selection of 100 points on a sphere of radius 3 around the origin. GMRES was restarted every 1500 iterations, with a stopping criterion of $\|\boldsymbol{r}\|_2 \leq 10^{-8}$. Copyright ©2019 Society for Industrial and Applied Mathematics. Reprinted with permission. All rights reserved

5.1 Mie Scattering

First, we test the implementation via the computation of the surface current induced by a plane wave from a perfectly conducting unit sphere. The data and some of the figures presented within this section were first published in [4].

An analytic solution to the density is known in terms of a series expansion, see [5] for a comprehensive account. Since the energy norm $\|\cdot\|_{\boldsymbol{H}_\times^{-1/2}(\mathrm{div}_\Gamma,\Gamma)}$ of the density is not computable explicitly, we choose to compare the $\boldsymbol{L}^2(\Gamma)$-error of the density. In accordance to quasi-optimality of the approach, cf. Theorem 3.48, a convergence of order p is expected, cf. Remark 3.29. As can be seen in Fig. 5.1a, the implementation behaves as predicted.

5.2 The Electric Field as a Quantity of Interest

Although the density obtained in an approach via the electric field integral equation admits a physical interpretation as the surface current, the quantity of interest of scattering problems is mainly the scattered electric field. This section will present corresponding numerical examples. The data was first published in [4].

In order to obtain a reference solution, we employ an approach via manufactured solution, cf. [6, Chap. 6.3], i.e., a function that fulfills the electric wave equation in Ω^c is used to generate the required Dirichlet data. By existence and uniqueness of the solution to Problem 2.31, cf. [7], one can validate the numerical scheme, i.e., an implementation to solve the discrete Problem 3.42.

As such a manufactured solution, we utilise a simple Hertz-dipole, for which one can check that it fulfils both (2.27) as well as Assumption 2.30.

Definition 5.1 (*Hertz-dipole,* [8, p. 411, (9.18)]). Let $\boldsymbol{x}_0 \in \Omega$. We define the function

$$\mathbf{DP}_{\boldsymbol{x}_0}(\boldsymbol{x}) := e^{-i\kappa r}\left(\frac{\kappa^2}{r}(\boldsymbol{n}\times\boldsymbol{p}_0)\times\boldsymbol{n} + \left(\frac{1}{r^3}-\frac{i\kappa}{r^2}\right)\left(3\boldsymbol{n}(\boldsymbol{n}\cdot\boldsymbol{p}_0)-\boldsymbol{p}_0\right)\right),$$

with $r = \|\boldsymbol{x}-\boldsymbol{x}_0\|_2$, $\boldsymbol{p}_0 = (0, 0.1, 0.1)$, and $\boldsymbol{n} = (\boldsymbol{x}-\boldsymbol{x}_0)/r$.

Given a reference solution, the errors illustrated in Fig. 5.1b validate the convergence rates of the electric field predicted by Corollary 3.49, where the maximal pointwise error in this and all upcoming tests is given as

$$\max_{\boldsymbol{x}\in V}\left\|\mathbf{DP}_{\boldsymbol{x}_0}(\boldsymbol{x}) - \boldsymbol{e}_h(\boldsymbol{x})\right\|_{\mathbb{C}^3}$$

for a point set V around the geometry, which will be further specified for each example.

The last data point of the highest order does not match the predicted order, but is, with an error around 10^{-12}, where numerical inaccuracies are to be expected.

Since the sphere-example is a classical benchmark test, we choose to publish detailed data about the computation, specifically in terms of time to solution, in Table 5.1. There, one can also find detailed information about the machine used for computations. This may serve as a reference to compare the presented approach to other implementations, but we stress again that one has to act cautiously when comparing times, since the performance of the fast method depends on various parameters of the problem, in particular, the product of the wavenumber κ to the size of the geometry. The input parameters of all computations are detailed in the captions of the corresponding figures.

Table 5.1 also makes clear that the conditioning of the system poses a problem, as is well-known and a field of research for the EFIE [9, 10]. One can clearly see that the time spent solving the system dominates the time to solution over the highly optimised assembly routines. This difference in time is visualised in Fig. 5.2.

While for classical lowest order elements quasi-optimal preconditioning strategies exist, see e.g. [10], a satisfying preconditioning strategy for the isogeometric approach is still an open area of research.

5.3 A Benchmark: Scattered Fields Around TESLA Cavities

To test more involved geometries with larger numbers of degrees of freedom, we test our boundary element method on the TESLA cavity geometries [4, 11]. They resemble the cavities as used in particle accelerators, as explained in the introduction of this thesis. It should be remarked that all of these examples avoid resonant

Table 5.1 Detailed data of the Sphere-example with $\kappa = 1$ and $\eta = 1.6$. Computed on a Workstation with Intel(R) Xeon(R) CPU E5-2670 0 @ 2.60 GHz, and has been compiled with `g++ 5.4`, with compile flags `-O3 -march=native -fopenmp`. The Mie-error refers to the error w.r.t. the analytic solution of the scattering problem described in Sect. 5.1, while the **DP**-error refers to the error obtained via the manufactured solution as described in Sect. 5.2, cf. Fig. 5.1a and b. Every 1500 iterations, the GMRES was restarted, with a stopping criterion of $\|\boldsymbol{r}\|_2 \leq 10^{-8}$. The **DP**-error was evaluated on a set of 100 points scattered across the sphere of radius 3. Copyright ©2019 Society for Industrial and Applied Mathematics. Reprinted with permission. All rights reserved

	$p = 1$			
h w.r.t. □	0.5	0.25	0.125	0.06125
Complex DOFs	48	192	768	3057
Matrix ass. (s)	0.02	0.14	1.14	9.15
Solving (s)	0.02	0.32	3.4	79.9
GMRES iterations	12	55	119	231
DP-error	0.0074	0.0009	0.0001	1.23e-05
Mie error ($\boldsymbol{L}^2$)	1.051	0.499	0.246	0.122
	$p = 2$			
h w.r.t. □	0.5	0.25	0.125	0.06125
Complex DOFs	108	300	972	3468
Matrix ass. (s)	0.06	0.55	4.8	47.3
Solving (s)	0.046	2.6	100.6	2279.6
GMRES iterations	48	158	362	616
DP-error	0.0009	1.82e-05	4.41e-07	1.29e-08
Mie error ($\boldsymbol{L}^2$)	0.251	0.052	0.012	0.0029
	$p = 3$			
h w.r.t. □	0.5	0.25	0.125	0.06125
Complex DOFs	192	432	1200	3888
Matrix ass. (s)	0.8	1.16	17.4	197.3
Solving (s)	0.15	8.46	237.8	8433
GMRES iterations	123	294	702	2003
DP-error	5.29e-05	9.83e-07	3.72e-09	2.45e-11
Mie error ($\boldsymbol{L}^2$)	0.085	0.011	0.0010	0.000121
	$p = 4$			
h w.r.t. □	0.5	0.25	0.125	0.06125
Complex DOFs	300	588	1452	4332
Matrix ass. (s)	0.6	5.42	52.1	746.29
Solving (s)	2.08	79.2	3072.9	78508
GMRES iterations	224	400	919	5681
DP-error	6.81-e06	1.54e-07	6.77e-11	8.33e-12
Mie error ($\boldsymbol{L}^2$)	0.021	0.0034	0.00012	6.69e-06

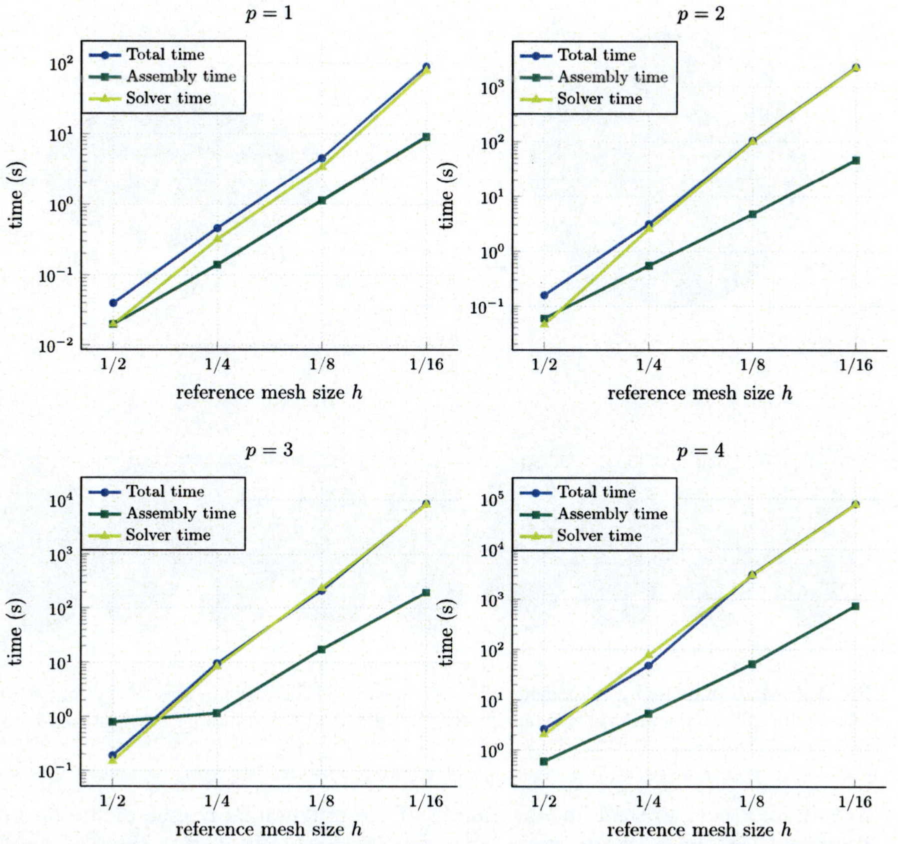

Fig. 5.2 Time to solution visualised for the Sphere-example. One can clearly see that the time spent solving the system dominates the time to solution over the highly optimised assembly routines. The data corresponds to that of Table 5.1. Copyright ©2019 Society for Industrial and Applied Mathematics. Reprinted with permission. All rights reserved

frequencies, and are only to be considered as a benchmark for solving scattering problems and to verify that the boundary element method works as desired. With cavities one is usually interested in solving the eigenvalue problem, which will be investigated in Chap. 6.

We start these numerical experiments on a single cell of the TESLA cavity, as depicted in Fig. 5.3, which resembles a single cell of the full nine-cell cavity. A volumetric discretisation is freely available through the `geopdes` package of Octave [12]. The scattering problem is then solved with a right hand side induced by the dipole for which the precise parameters are presented in Fig. 5.4 and Table 5.2. The results are depicted in Fig. 5.4.

One can still observe the high convergence rates w.r.t. the number of degrees of freedom. One can also see that the time for matrix assembly, as well as the time

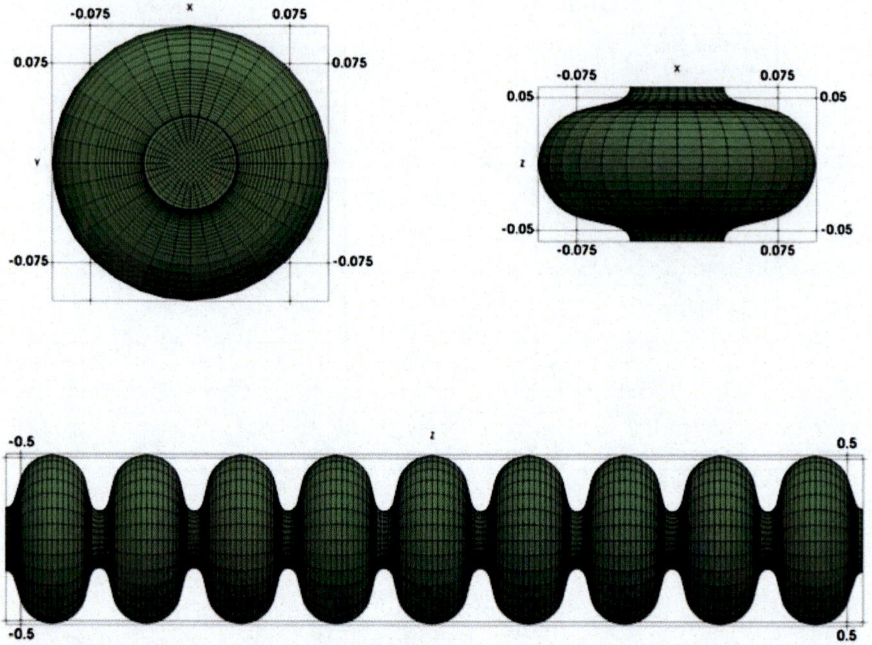

Fig. 5.3 Mesh induced by refinement of level 3 of the TESLA geometries. Copyright ©2019 Society for Industrial and Applied Mathematics. Reprinted with permission. All rights reserved

to solution, seem to scale independently of the polynomial degree of the discrete functions. However, both the time for matrix assembly and the time to solution differ by a constant factor, favoring solutions obtained via higher-order approaches. Moreover, one observes that the number of GMRES iterations required for the solution of the system w.r.t. the achieved accuracy of the solution scales independently of p, suggesting that the iteration count to solve the system remains constant for a desired accuracy, weather this accuracy is achieved via p or h refinement. This slightly favors higher-order approaches: For a set accuracy threshold, systems of higher order approaches are smaller due to the higher accuracy per DOF. Thus, each iteration of a matrix-free solver is computationally cheaper.

For the nine-cell example, such clear behavior is not visible, cf. Table 5.2. We attribute this to the fact that the compression parameters (η for admissibility condition and order q of the multipole interpolation) had to be chosen such that the problem remained computable on the accessible machines, i.e., one can not depend on the result of Theorem 4.8. Despite the suboptimal choice of parameters η and q, one still can observe that the method converges and yields good results.

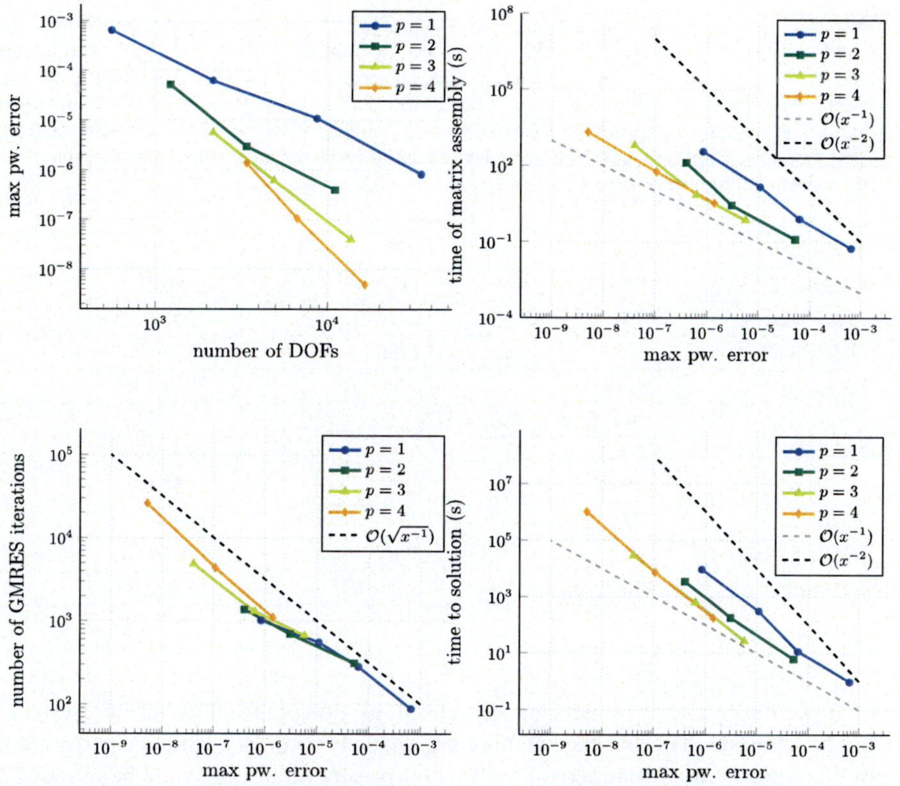

Fig. 5.4 Results for the TESLA one-cell geometry. Wavenumber $\kappa = 18$, manufactured solution $\mathbf{DP}_{(0,0.1,0.1)}$. Admissibility condition with $\eta = 0.1$ and $q = 14$. GMRES restart after 1500 iterations, stopping criterion $\|\boldsymbol{r}\|_2 \leq 10^{-10}$. The **DP**-error refers to the maximum error obtained via the manufactured solution of a selection of 100 points on a sphere of radius 3 around the origin. Copyright ©2019 Society for Industrial and Applied Mathematics. Reprinted with permission. All rights reserved

5.4 Comparison to Raviart–Thomas Elements

This section compares the isogeometric basis $\mathbb{S}^1_{\boldsymbol{p},\boldsymbol{\Xi}}(\Gamma)$ to the parametric quadrilateral Raviart–Thomas $\mathbb{Q}_p$ of elements $\boldsymbol{Q}_p$, where each element is of the form $\boldsymbol{Q}_p = Q_{p,p-1} \times Q_{p-1,p}$, for $Q_{p_1,p_2} = \mathbb{P}_{p_1} \otimes \mathbb{P}_{p_2}$, cf. [13, Sect. 5.2.2].

We define the elements $\boldsymbol{Q}_p$ on the isogeometric elements induced by the knot vectors in reference domain and then apply the push-forward of the geometry mapping to get a discrete space in physical domain. Note that on $\boldsymbol{Q}_p$ normal continuity must be enforced as well. As with the isogeometric discretisation, this is implemented by means of a transformation as in (4.6).

Table 5.2 Detailed data of the nine-cell TESLA example with $\kappa = 10$. Computed on a Workstation, with Intel(R) Xeon(R) CPU E7- 8850, and has been compiled with `g++ 4.8.5`, with compile flags `-O3 -march=native -fopenmp`. The **DP**-error refers to the maximum error obtained via the manufactured solution of a selection of 100 points on a sphere of radius 3 around the origin. The stopping criterion for the GMRES was a residual of $\|\boldsymbol{r}\|_2 < 10^{-10}$, with a restart every 1500 iterations. Copyright ©2019 Society for Industrial and Applied Mathematics. Reprinted with permission. All rights reserved

	$p = 2, q = 12, \eta = 0.15$		
h w.r.t. □	0.5	0.25	0.125
Complex DOFs	4068	11300	36612
Matrix ass. (s)	43	64	2031
GMRES iterations	879	1230	2552
DP-error	1.69e-03	3.84e-07	9.79e-09
	$p = 3, q = 10, \eta = 0.3$		
h w.r.t. □	0.5	0.25	0.125
Complex DOFs	7232	16272	45200
Matrix ass. (s)	37	207	5944
GMRES iterations	1424	2987	7934
DP-error	9.07e-07	3.28e-07	1.33e-09

Since the geometry transformation is used, we do not introduce meshing errors in our benchmarks. This means that this comparison is optimistic w.r.t. the performance of $\boldsymbol{Q}_p$ since in most commercial tools a non-parametric mesh would be introduced.

We remark that by this construction, the lowest order spline space coincides with the lowest order Raviart–Thomas space. This can be observed the results of all experiments, cf. Figs. 5.5, 5.6, and 5.9. In addition to the visualisations, detailed numerical data of selected simulations are depicted in Table 5.3. All data and some figures presented within this section were first published in [14].[1]

5.4.1 *The Unit Sphere*

We start our comparison with the unit sphere. The dipole is positioned at $(0.1, 0.1, 0)^\top$ and the wavenumber chosen to be $\kappa = 1$. The error measurement corresponds to the maximum pointwise error (in the complex Cartesian sense), evaluated at 100 points on a sphere with radius 3 around the origin. Due to the smooth geometry the effect of higher-order basis functions w.r.t. the order of convergence of $\mathcal{O}(h^{2p+1})$ is clearly visible, see Fig. 5.5.

---

[1] ©2019, IEEE. Reprinted, with permission, from Dölz et al., "A Numerical Comparison of an Isogeometric and a Parametric Higher Order Raviart–Thomas Approach to the Electric Field Integral Equation", Jan 2020.

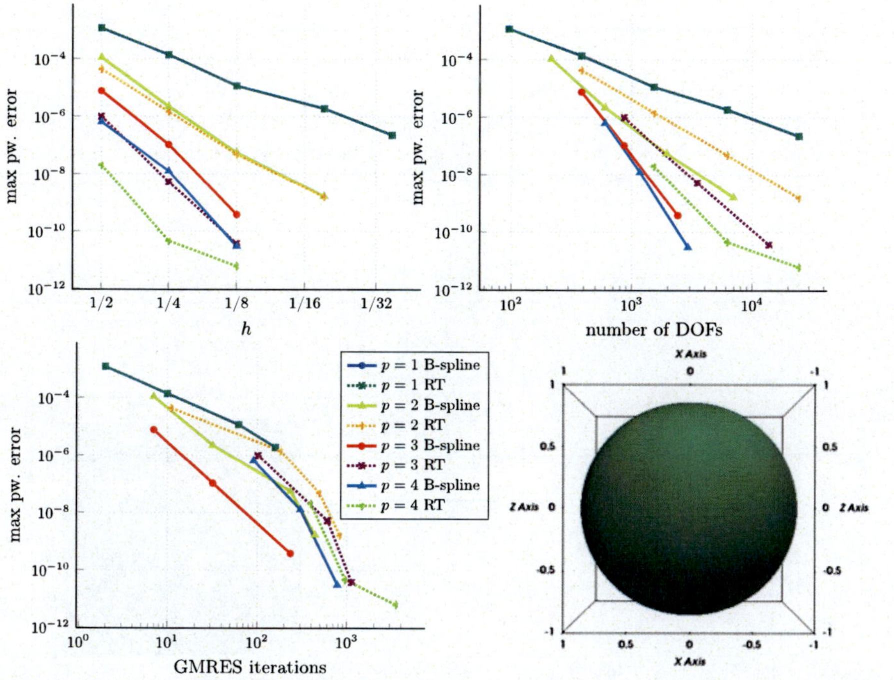

Fig. 5.5 Sphere-example, $\boldsymbol{x}_0 = (0.1, 0.1, 0)^\top$, $\kappa = 1$. The isogeometric approach shows to be beneficial w.r.t. accuracy per degree of freedom. Copyright ©2020 IEEE. Reprinted, with permission, from [14]

For the same level of refinement the Raviart–Thomas discretisation yields better accuracies. This is expected, since for the same polynomial degrees and the same level of refinement, the spline discretisation is contained in the corresponding Raviart–Thomas space. In contrast, the B-spline spaces yield an improved accuracy per degree of freedom (DOF) as visualised in Fig. 5.5.

5.4.2 *The Fichera Cube*

We now employ a 24 patch Fichera cube with $\boldsymbol{x}_0 = (0.5, 0.5, 0.5)^\top$ and $\kappa = 1$ as a test case. The choice of evaluation points coincide with the previous example.

As with the sphere, the Raviart–Thomas elements yield an increased accuracy w.r.t. the same refinement level, whereas the B-splines admit an improved accuracy per DOF, cf. Fig. 5.6.

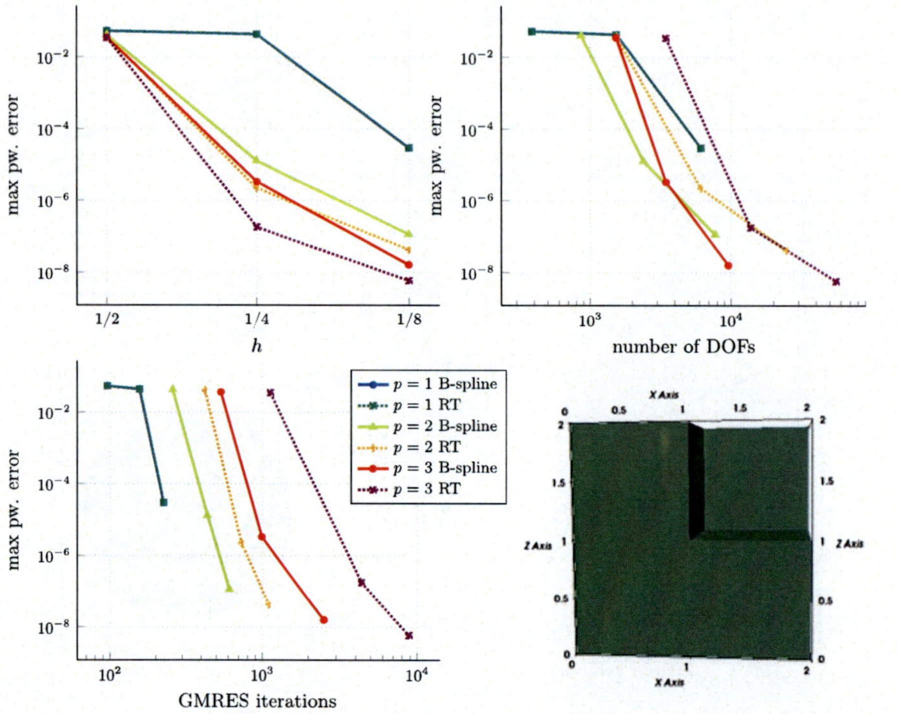

Fig. 5.6 Fichera-example. Dipole placed at $\boldsymbol{x}_0 = (0.5, 0.5, 0.5)^\top$ with wavenumber $\kappa = 1$. This example admits improved accuracy per degree of freedom compared to the Raviart–Thomas elements, as well. Copyright ©2020 IEEE. Reprinted, with permission, from [14]

5.4.3 The Ship Geometry

As a last example of this comparison, we test with the ship geometry depicted in Fig. 5.7. The mesh underlying the second refinement level can be seen in Fig. 5.8. The ship is mad up of 28 quadratic NURBS patches. This time we solve an interior problem with a dipole placed at $\boldsymbol{x}_0 = (7, 2, 0)^\top$ and the wavenumber chosen to be $\kappa = 5$. The accuracy of the solution is measured at a set of nine points around the point $(1, 0, 0)^\top$.

In analogy to the previous tests, the Raviart–Thomas space achieves higher accuracies if one compares the same underlying mesh, i.e., the element structure induced by the level of refinement, cf. Fig. 5.9, and in terms of accuracy per DOF the spline space is superior.

Table 5.3 Showcase of examples of comparable accuracy. The simulation was stopped once the specified accuracy w.r.t. the manufactured solution was achieved. Copyright ©2020 IEEE. Reprinted, with permission, from [14]

		Example 1	Example 2	Example 3
Geometry		Sphere	Fichera	Ship
p		4	3	2
Refinement level	Splines	3	3	4
	Raviart–Thomas	2	2	4
Number of DOFs	Splines	2904	9600	32368
	Raviart–Thomas	6144	13824	114688
Accuracy spec.		1e-10	1e-06	1e-04
Error	Splines	3.021e-11	1.626e-08	5.724e-05
	Raviart–Thomas	4.663e-11	1.809e-07	3.747e-05
Iterations	Splines	783	2505	7483
	Raviart–Thomas	958	4363	58169

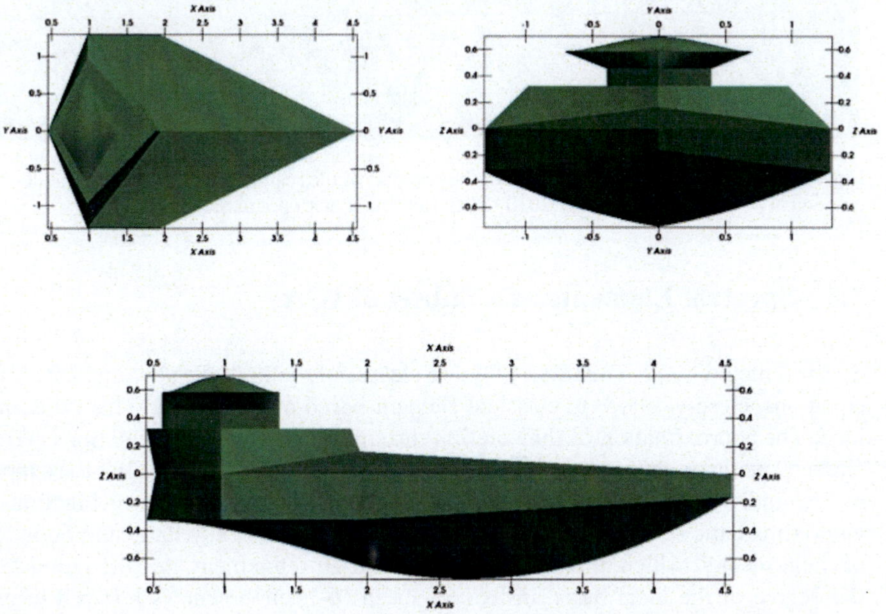

Fig. 5.7 The ship geometry. Copyright ©2020 IEEE. Reprinted, with permission, from [14]

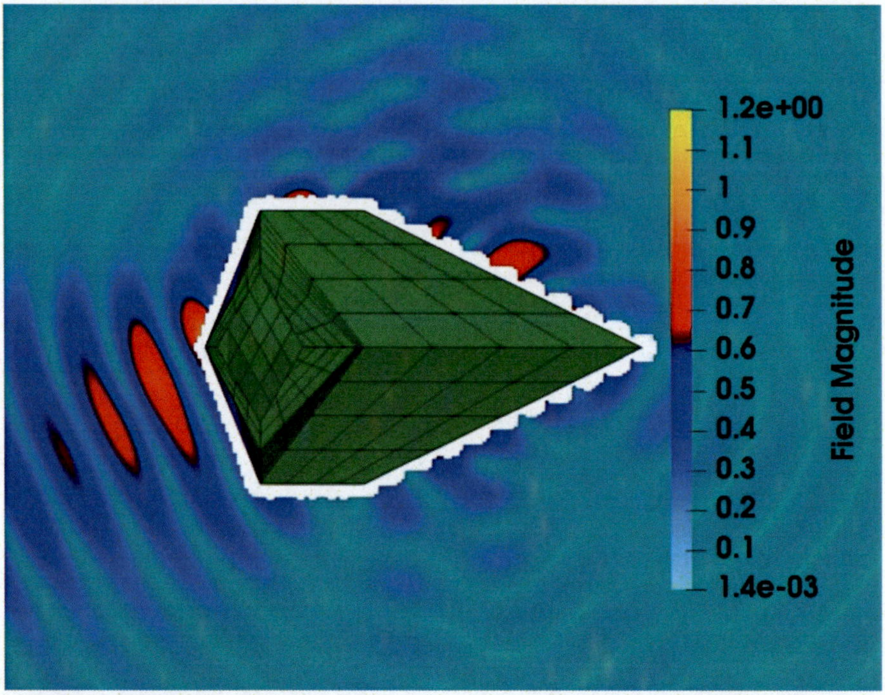

Fig. 5.8 Ship geometry and magnitude of the real part of the scattered field with mesh induced by refinement of level 2. Copyright ©2020 IEEE. Reprinted, with permission, from [14]

5.5 Spectral Elements as a Subset of IGA

Spectral methods, c.f. Trefethen [15], are classes of methods for solving differential equations, closely related to classical element-based methods. They share the same idea: The approximation of the solution through a series of basis functions. While classical methods choose to refine a mesh, and with this to further localise the support of each individual basis function, spectral methods employ global basis functions to approximate the solution of the problem. Thus, quite often any finite and boundary element method which relies on (possibly local) p-refinement, i.e., the increase of the degree of the local basis, instead of mesh (h-) refinement, is referred to as a *spectral element method*.

In engineering applications, spectral element methods are rarely considered; the reason simply being that to fully enjoy their convergence properties, meshes with curved elements of increasing orders must be generated. This poses challenges to mesh generation and pre-processing. In contrast, classical h-refinement based mesh generators are well understood. This caveat can be circumvented through the utilisation of parametric mappings as in the case of isogeometric analysis.

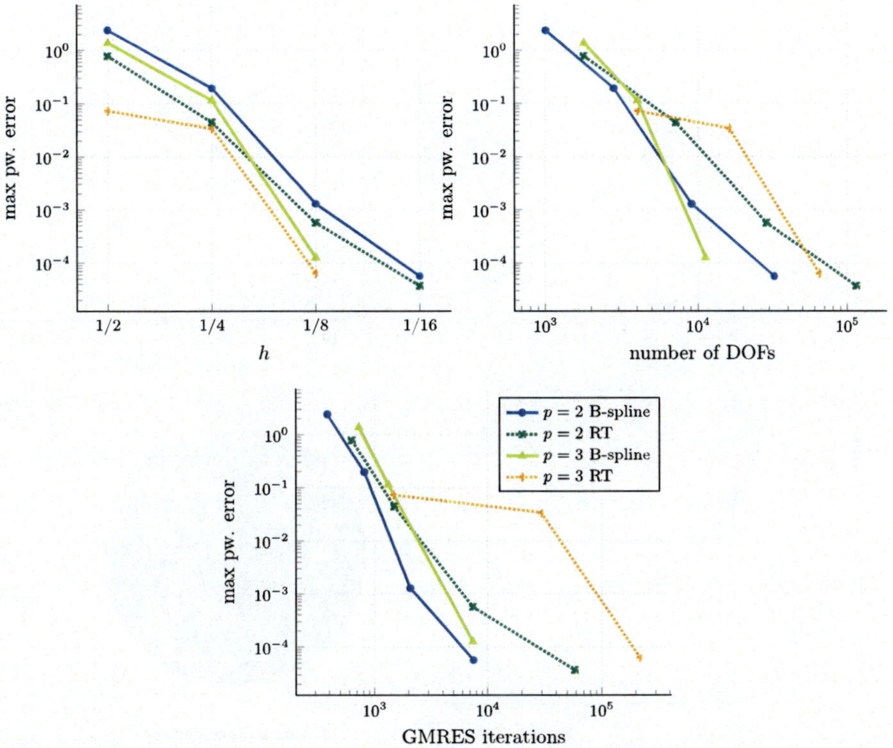

Fig. 5.9 Ship example, $\boldsymbol{x}_0 = (7, 2, 0)^\top$, $\kappa = 5$. Copyright ©2020 IEEE. Reprinted, with permission, from [14]

This section showcases examples of the solution of electromagnetic scattering problems through p-refinement, whose data was first published in [1]. We employ p-refinement to patchwise Bernstein polynomials. We remark that a dedicated implementation should use basis better suited for spectral approaches, cf. [15], and that the Bernstein polynomials were chosen for mere convenience.

As a first example, we choose the example of a unit sphere and define a dipole with $\boldsymbol{x}_0 = \boldsymbol{p}_0 = (0, 0.1, 0.1)$ as an excitation. Although a boundary element framework has been utilised to solve the problem, no compression was applied to the systems due to their small system size, cf. Fig. 5.10. Thus, the condition of the system matters little compared to the case in which compression must be applied, since direct solvers can be applied. In this case, the partially pivoted LU decomposition of [16] was used.

On the sphere, a stable exponential rate of convergence w.r.t. p is observed. The application of p-refinement reduces the time required for matrix assembly as well as the overall system size, cf. Fig. 5.10, significantly (Fig. 5.11).

As a second geometry, we investigate the boat. Placing the dipole inside with $\boldsymbol{x}_0 = (1, 0, 0)$ and $\boldsymbol{p}_0 = (0, 0, 0.1)$, we evaluate the electric field around the geometry. The rate of convergence is not as pronounced as before. Still, one can see an exponential convergence behaviour paired with excellent times to solution.

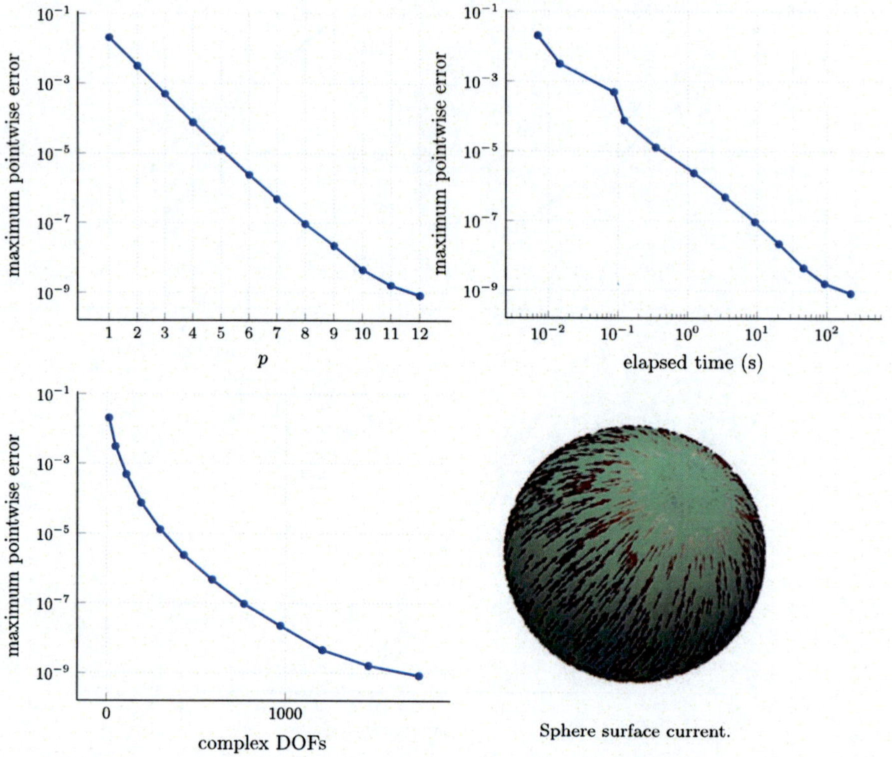

Fig. 5.10 Numerical examples on the unit sphere. Wavenumber $\kappa = 1$. The error is evaluated on a selection of 10 000 points on a sphere of radius 2 around the origin. The computation was conducted on a desktop PC with 16 GB RAM and an Intel i7 8600k processor

Both examples show that p-refinement alone is a valid alternative to h-refinement. Moreover, as mentioned before, the reduced system size allows for the application of direct solvers. However, since with higher p the conditioning of the system deprecates rapidly, one will eventually reach the point where even direct solvers start to struggle. This might explain the declining convergence rate observed in Fig. 5.10 for $p = 11, 12$, and could possibly be avoided by the utilisation of a different polynomial basis.

5.6 Compression Rates

To conclude this chapter, we will briefly mention compression rates. To verify the asymptotic behaviour of the compression scheme, the memory which would be allocated by the implementation if the corresponding matrices were to be assembled has been recorded. The visualised data can be seen in Fig. 5.12.

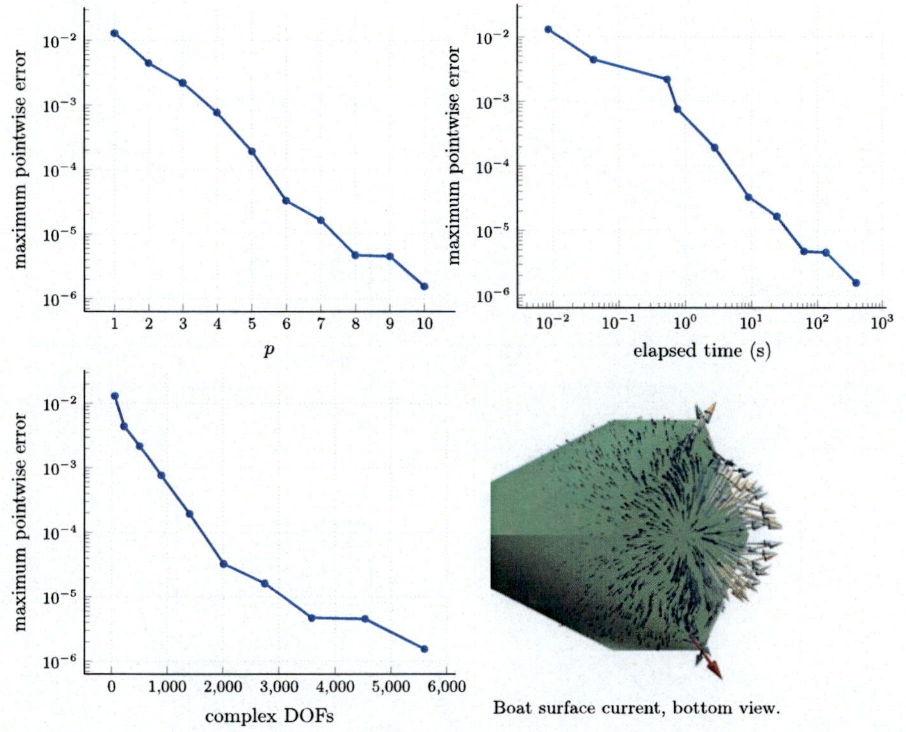

Fig. 5.11 Numerical examples for the ship geometry. The highest value of p generates a discretisation with 5 600 complex-valued degrees of freedom. Wavenumber $\kappa = 1$. The error was evaluated on a selection of 10 000 points on a sphere of radius 6 around the origin. The computation was conducted on a desktop PC with 16 GB RAM and an Intel i7 8600k processor

Therein, "$\boldsymbol{V}_{\kappa,h}$ uncompressed" denotes the matrix $\boldsymbol{V}_{\kappa,h}$ corresponding to the left hand side of (4.6). "$\boldsymbol{V}^*_{\kappa,h}$ uncompressed" denotes the local element matrix assembled w.r.t. the superspace $\mathbb{S}^*_{p,m}$, and "$\boldsymbol{V}^*_{\kappa,h}$ compressed" refers to the local element matrix after application of the interpolation-based multipole scheme.

Due to the superspace approach and the fact that the compression scheme can only be effectively applied to larger levels of refinement all of these examples show a range where a dense assembly of the small system is favourable. However, for larger levels of refinement, i.e., more degrees of freedom, the compression proves to be effective. All examples show linear compression behaviour, since the interpolation degree q of the multipole method has been fixed. Note that, to maintain asymptotically optimal behaviour, q should be chosen to scale logarithmically with the number of degrees of freedom, see Theorem 4.8. However, all previous numerical examples of the sphere and the one-cell TESLA geometry were conducted with the same value of $q = 10$ as in Fig. 5.12. The choice to fix q as a constant and sufficiently high parameter was made deliberately to simplify reproduction and parameter dependence of the numerical examples.

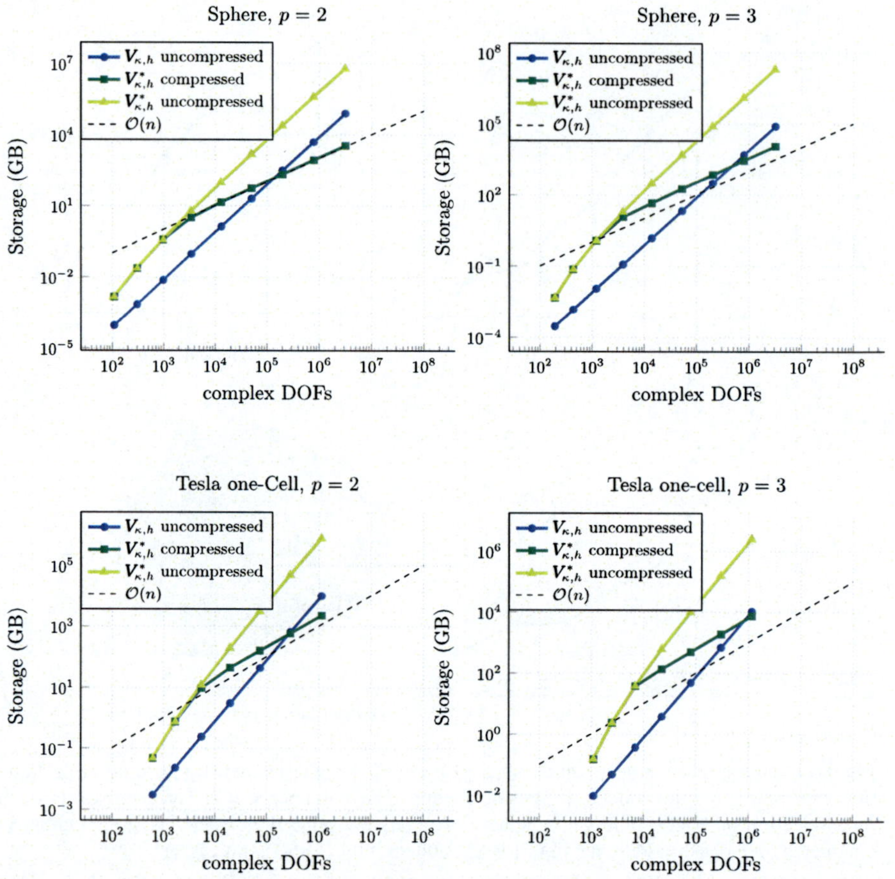

Fig. 5.12 Efficiency of the compression scheme and storage requirements. For both examples $\eta = 1.6$ was set. For the sphere the FMM-degree $q = 8$ was chosen, while for the Cavity a fixed $q = 10$ was utilised

It must also be remarked that the interpolation-based multipole method does not only reduce the storage requirements, but also the complexity of the matrix-vector product, cf. Theorem 4.8.

5.7 Summary of the Numerical Experiments

We discussed multiple numerical experiments showcasing that the proposed isogeometric discretisation of the electric field integral equation, i.e., our numerical scheme to Problem 3.42, is a valid alternative to the other methods briefly introduced within the introduction.

First, a classical Mie-scattering example in Sect. 5.1 verified the quasi-optimality of the unknown surface current as predicted by Theorem 3.48. In Sect. 5.2 and the following, pointwise numerical evaluations of the representation in Corollary 2.36 were shown to behave as predicted by Corollary 3.49. We furthermore compared our approach to other discretisations w.r.t. their accuracy per degree of freedom. Specifically, the doubled convergence order of $2p + 1$ w.r.t. the error in the electric field was verified through an approach via manufactured solutions.

Afterwards, in Sect. 5.5, an approach via p-refinement was applied successfully. Moreover, the asymptotic behaviour of the storage requirements was verified through the implementation as well, which was previously predicted by Theorem 4.8.

However, the numerical experiments also make clear that the development of a preconditioning strategy for the isogeometric EFIE is an important open topic of research.

References

1. Kurz S, Schöps S, Wolf F (2019) Towards a spectral method of moments using computer aided design
2. OpenMP Architecture Review Board. OpenMP Application Program Interface Version 3.0. May 2008
3. Dölz J, Harbrecht H, Kurz S, Multerer M, Schöps S, Wolf F (2019) Bembel. The C++ boundary element engineering library. Official website, www.bembel.eu. Accessed 20 Aug 2019
4. Dölz J, Kurz S, Schöps S, Wolf F (2019) Isogeometric boundary elements in electromagnetism: rigorous analysis, fast methods, and examples. SIAM J Sci Comput 41(5):B983–B1010
5. Weggler L (2011) High order boundary element methods. Dissertation. Universitat des Saarlandes, Saarbrücken
6. Oberkampf WL, Roy CJ (2010) Verification and validation in scientific computing. Cambridge University Press, New York
7. Buffa A, Christiansen S (2003) The electric field integral equation on Lipschitz screens: definitions and numerical approximation. Numerische Mathematik 94(2):229–267
8. Jackson JD (1998) Classical Electrodynamics, 3rd edn. Wiley, New York
9. Christiansen SH, Nédélec J-C (2002) A preconditioner for the electric field integral equation based on Calderon formulas. SIAM J Numer Anal 40(3):1100–1135
10. Andriulli FP, Cools K, Bagci H, Olyslager F, Buffa A, Christiansen S, Michielssen E (2008) A multiplicative Calderon preconditioner for the electric field integral equation. IEEE Trans Antennas Propag 56(8):2398–2412
11. Aune B, Bandelmann R, Bloess D, Bonin B, Bosotti A, Champion M, Crawford C, Deppe G, Dwersteg B, Edwards DA, Edwards HT, Ferrario M, Fouaidy M, Gall P-D, Gamp A, Gössel A, Graber J, Hubert D, Hüning M, Juillard M, Junquera T, Kaiser H, Kreps G, Kuchnir M, Lange R, Leenen M, Liepe M, Lilje L, Matheisen A, Möller W-D, Mosnier A, Padamsee H, Pagani C, Pekeler M, Peters H-B, Peters O, Proch D, Rehlich K, Reschke D, Safa H, Schilcher T, Schmuser P, Sekutowicz J, Simrock S, Singer W, Tigner M, Trines D, Twarowski K, Weichert G, Weisend J, Wojtkiewicz J, Wolff S, Zapfe K (2000) Superconducting TESLA cavities. Phys Rev Accel Beams 3(9):092001
12. de Falco C, Reali A, Vázquez R (2011) GeoPDEs: a research tool for isogeometric analysis of PDEs. Adv Eng Softw 42:1020–1034
13. Zaglmayr S (2006) High order finite element methods for electromagnetic field computation. Dissertation. Johannes Kepler Universität, Linz, Austria

14. Dölz J, Kurz S, Schöps S, Wolf F (2019) A numerical comparison of an isogeometric and a parametric higher order Raviart-Thomas approach to the electric field integral equation. IEEE Trans Antennas Propag 68(1):593–597
15. Trefethen L (2000) Spectral methods in MATLAB. SIAM, Philadelphia
16. Benoît J, Guennebaud G (2019) Eigen3 C++ linear algebra template library. Official website, https://eigen.tuxfamily.org. Accessed 20 Aug 2019

Chapter 6
The Discrete Eigenvalue Problem

This chapter is devoted to the discussion of the solution of Problem 2.32, i.e., the computation of resonant frequencies within perfectly conducting structures.

Noting that the perfect electric conductor boundary condition, i.e., Assumption 2.29, requires the solution $\boldsymbol{E}$ of the eigenvalue problem to fulfill $\boldsymbol{\gamma}_t(E) = 0$, the single layer representation of Corollary 2.36 immediately yields the following discrete counterpart to Problem 2.32.

Problem 6.1 (Discrete Eigenvalue Problem) *Find* $\boldsymbol{w}_h \in \mathbb{S}^1_{p,\boldsymbol{\Xi}}(\Gamma)$ *and* $\kappa \in (0, \infty)$ *such that*

$$\langle \mathbf{V}_\kappa \boldsymbol{w}_h, \boldsymbol{\mu}_h \rangle_\times = 0, \tag{6.1}$$

for all $\boldsymbol{\mu}_h \in \mathbb{S}^1_{p,\boldsymbol{\Xi}}(\Gamma)$.

As mentioned in the introduction of this thesis, the solution of this problem, specifically within the design of accelerator cavities, requires high accuracy. While the boundary element method promises superb convergence behaviour, cf. Remark 3.50, the non-linear dependency on the wavenumber within the fundamental solution u^*_κ renders the linear eigenvalue problem stated in Problem 2.32 non-linear.

However, the recently introduced class of contour integral methods offers a solution to this caveat.

This chapter consists of two parts. First, in Sect. 6.1 we introduce the contour integral method. The second part, Sect. 6.2, discusses numerical examples showcasing the solution of Problem 6.1 via the combination of a contour integral method and our proposed isogeometric boundary element method.

© The Author(s), under exclusive license to Springer Nature Switzerland AG 2021
F. Wolf, *Analysis and Implementation of Isogeometric Boundary Elements for Electromagnetism*, Springer Theses,
https://doi.org/10.1007/978-3-030-61939-8_6

6.1 The Contour Integral Method

This section will give a short summary of the contour integral method as introduced by Beyn [1], without any non-trivial modifications. Note that there are alternative approaches by other authors, where the first publication seems to go back to Asakura et al. [2], compare also [3]. After this short review, we will discuss three numerical examples that utilise this method to solve Problem 2.32. Note that for this entire section m will refer to the dimension of the discrete space of the proposed boundary element method instead of the refinement level.

Let $\boldsymbol{T}: D \to \mathbb{C}^{m\times m}$ be holomorphic on some compact and connected domain $D \subset \mathbb{C}$. By $\boldsymbol{T}^{\mathrm{h}}$ we denote the Hermitian transpose of $\boldsymbol{T}$, and $\boldsymbol{T}^{\mathsf{T}}$ denotes its usual transpose. We want to solve a non-linear eigenvalue problem

$$\boldsymbol{T}(z)\boldsymbol{v} = 0, \quad \boldsymbol{v} \in \mathbb{C}^m, \quad \boldsymbol{v} \neq \boldsymbol{0}, \quad z \in D.$$

If z is not an eigenvalue of $\boldsymbol{T}$, we find by the condition above that the kernel of $\boldsymbol{T}(z)$ is trivial, thus $\boldsymbol{T}(z)$ has full rank and is invertible.

For the remainder of this chapter, we set $\widehat{\Gamma} = \partial D$. Assume there to be k eigenvalues $(\lambda_j)_{j\leq k}$ in the interior of D, such that all eigenvectors are linearly independent. We remark that the entire approach of Beyn [1] was also generalised to the case of linearly dependent eigenvectors.

As a consequence of the version of the Keldysh Theorem as proven by Mennicken and Möller [4], we can provide the following generalisation of the classical residue theorem.

Theorem 6.2 (Contour Integral Theorem, [1, Theorem 2.9.]) *For T as above and holomorphic $f: D \to \mathbb{C}$, it holds that*

$$\frac{1}{2\pi i}\int_{\widehat{\Gamma}} f(z)\boldsymbol{T}(z)^{-1}\,\mathrm{d}z = \sum_{j=1}^{k} f(\lambda_j)\boldsymbol{v}_j\boldsymbol{w}_j^{\mathrm{h}},$$

where $\boldsymbol{v}_j$ and $\boldsymbol{w}_j$ are left and right eigenvectors corresponding to λ_n that are normalized according to

$$\boldsymbol{w}_j^{\mathrm{h}}\boldsymbol{T}^{\mathsf{T}}(\lambda_j)\boldsymbol{v}_j = 1$$

for all $j \leq k$.

Note that the left hand side of Problem 6.1 can be interpreted as a matrix-valued function $\boldsymbol{X}: D \to \mathbb{C}^{m\times m}$ with $\boldsymbol{X}(\kappa) = \mathbf{V}_{\kappa,h}$, where $\mathbf{V}_{\kappa,h}$ is defined in (3.52). Thus the contour integral method can use Theorem 6.2 to find a solution to Problem 6.1 as follows.

Theorem 6.3 (Linearisation of Eigenvalue Problems, [1, Theorem 3.1]) *Suppose that $T: D \to \mathbb{C}^{m\times m}$ is holomorphic and the eigenvectors of all eigenvalues $(\lambda_j)_{0\leq j<k}$ in some compact and connected domain D are linearly independent. Then, there*

exists a diagonalisable matrix $\boldsymbol{B}$ *which can be computed from* $\boldsymbol{X}$ *such that* $\boldsymbol{B}$ *that has the same eigenvalues as the eigenvalue problem under consideration within D.*

Proof Let $\boldsymbol{V}, \boldsymbol{W} \in \mathbb{C}^{m\times k}$ be the matrices

$$\boldsymbol{V} = (\boldsymbol{v}_1 \cdots \boldsymbol{v}_k)$$
$$\boldsymbol{W} = (\boldsymbol{w}_1 \cdots \boldsymbol{w}_k)$$

of left and right sided eigenvectors, respectively. Assume them to have full rank. We choose some random matrix $\widehat{\boldsymbol{V}} \in \mathbb{C}^{m\times \ell}$ for $k \leq \ell \leq m$ such that $\boldsymbol{W}^{\mathrm{h}}\widehat{\boldsymbol{V}} \in \mathbb{C}^{k\times \ell}$ has full rank.

One can check that $\boldsymbol{X}$ is holomorphic since the fundamental solution can be represented as a power series, compare also [5, Lemma 2.3]. We define the matrices

$$\boldsymbol{A}_0 = \frac{1}{2\pi i}\int_{\widehat{\Gamma}} \boldsymbol{X}(z)^{-1}\widehat{\boldsymbol{V}} \,\mathrm{d}z \in \mathbb{C}^{m\times \ell} \qquad \text{and} \tag{6.2}$$

$$\boldsymbol{A}_1 = \frac{1}{2\pi i}\int_{\widehat{\Gamma}} z\boldsymbol{X}(z)^{-1}\widehat{\boldsymbol{V}} \,\mathrm{d}z \in \mathbb{C}^{m\times \ell}, \tag{6.3}$$

and thus, by Theorem 6.2, obtain the identities

$$\boldsymbol{A}_0 = \sum_{j=1}^{k} \boldsymbol{v}_j \boldsymbol{w}_j^{\mathrm{h}} \widehat{\boldsymbol{V}} = \boldsymbol{V}\boldsymbol{W}^{\mathrm{h}}\widehat{\boldsymbol{V}} \qquad \text{and} \tag{6.4}$$

$$\boldsymbol{A}_1 = \sum_{j=1}^{k} \lambda_j \boldsymbol{v}_j \boldsymbol{w}_j^{\mathrm{h}} \widehat{\boldsymbol{V}} = \boldsymbol{V}\boldsymbol{\Lambda}\boldsymbol{W}^{\mathrm{h}}\widehat{\boldsymbol{V}}. \tag{6.5}$$

The matrix $\boldsymbol{\Lambda}$ is the diagonal matrix with the jth eigenvalue in the jth row.

By means of a singular value decomposition we find

$$\boldsymbol{A}_0 = \boldsymbol{V}\boldsymbol{W}^{\mathrm{h}}\widehat{\boldsymbol{V}} = \boldsymbol{V}_0\boldsymbol{\Sigma}_0\boldsymbol{W}_0^{\mathrm{h}} \tag{6.6}$$

where $\boldsymbol{V}_0$ is a $m \times k$ matrix, $\boldsymbol{W}_0$ is an $\ell \times k$ matrix, $\boldsymbol{V}_0^{\mathrm{h}}\boldsymbol{V}_0 = \boldsymbol{W}_0^{\mathrm{h}}\boldsymbol{W}_0 = \boldsymbol{I}$ and $\boldsymbol{\Sigma}_0$ is the diagonal matrix with (non zero) singular values $\sigma_1 \geq \cdots \geq \sigma_k > 0$.

It is clear that $\boldsymbol{V} = \boldsymbol{V}_0\boldsymbol{V}_0^{\mathrm{h}}\boldsymbol{V}$, and thus we set $\boldsymbol{S} := \boldsymbol{V}_0^{\mathrm{h}}\boldsymbol{V} \in \mathbb{C}^{k\times k}$ where $\boldsymbol{S}$ is invertible since $\boldsymbol{V}_0$ and $\boldsymbol{V}$ have maximal rank.

We can now conclude

$$\boldsymbol{V}_0\boldsymbol{S} = \boldsymbol{V}. \tag{6.7}$$

By (6.6) and (6.7) we get

$$\boldsymbol{V}_0\boldsymbol{S}\boldsymbol{W}^{\mathrm{h}}\widehat{\boldsymbol{V}} = \boldsymbol{V}_0\boldsymbol{\Sigma}_0\boldsymbol{W}_0^{\mathrm{h}}, \tag{6.8}$$

and since $\boldsymbol{V}_0$ has rank k we get

$$\boldsymbol{W}^{\mathrm{h}}\widehat{\boldsymbol{V}} = \boldsymbol{S}^{-1}\boldsymbol{\Sigma}_0\boldsymbol{W}_0^{\mathrm{h}}. \tag{6.9}$$

Inserting (6.9) in (6.5) together with the definition of $\boldsymbol{S}$ yields

$$\boldsymbol{V}_0^{\mathrm{h}}\boldsymbol{A}_1 = \boldsymbol{S\Lambda W}^{\mathrm{h}}\widehat{\boldsymbol{V}} = \boldsymbol{S\Lambda S}^{-1}\boldsymbol{\Sigma}_0\boldsymbol{W}_0^{\mathrm{h}}. \tag{6.10}$$

Since $\boldsymbol{\Sigma}_0$ is invertible and $\boldsymbol{W}_0$ is unitary we obtain

$$\boldsymbol{S\Lambda S}^{-1} = \boldsymbol{V}_0^{\mathrm{h}}\boldsymbol{A}_1\boldsymbol{W}_0\boldsymbol{\Sigma}_0^{-1}. \tag{6.11}$$

Thus, the eigenvalues of $\boldsymbol{B} := \boldsymbol{V}_0^{\mathrm{h}}\boldsymbol{A}_1\boldsymbol{W}_0\boldsymbol{\Sigma}_0^{-1}$ will be the eigenvalues of $\boldsymbol{X}$ within D. ■

The proof is constructive, which means we can immediately extract a working algorithm. In the following, this algorithm is outlined in human-readable form. It corresponds to the implemented method up to performance optimisations, as will be discussed below.

Note that the algorithm is capable of identifying eigenvalues of the generalised eigenvalue problem without evaluating at or close to those values. This means that it is applicable to a discretisation of the electric field integral equation without any problems, since an approach via EFIE becomes instable at resonant frequencies.

Algorithm (Contour Integral Method, [1, p. 15]). Let $\boldsymbol{X}$ be given as above, $\{\widehat{\Gamma}_n\}_{n\leq N}$ be a set of N equidistant quadrature points on some $\widehat{\Gamma} = \partial D$ as above, $\delta, \epsilon > 0$ and $\ell < m$.

Data: $\{\widehat{\Gamma}_n\}_{n\leq N}, \boldsymbol{X}, \delta, \epsilon, \ell$
Result: $(\lambda_j)_{j\leq k}$
kfound ← false;
$m \leftarrow \mathrm{size}(\boldsymbol{X})$;
while *kfound* == *false* **do**
  $\widehat{\boldsymbol{V}} \leftarrow \mathrm{RandomFullRank}(m \times \ell)$;
  $\boldsymbol{A}_0 \leftarrow \frac{1}{iN}\sum_{j=1}^{N} \boldsymbol{X}(\widehat{\Gamma}_j))^{-1}\widehat{\boldsymbol{V}}$;
  $(\boldsymbol{V}, \boldsymbol{\Sigma}, \boldsymbol{W}^{\mathrm{h}}) \leftarrow \mathrm{SVD}(\boldsymbol{A}_0)$; // $\boldsymbol{\Sigma} = \mathtt{diag}(\sigma_1, \ldots, \sigma_\ell)$
  $k \leftarrow j$, where $\sigma_1 \geq \cdots \geq \sigma_j > \delta > \sigma_{j+1} \approx \cdots \approx \sigma_\ell \approx 0$;
  if $k < \ell$ **then**
    kfound ← true;
    $\boldsymbol{V}_0 \leftarrow \boldsymbol{V}(1\colon m, 1\colon k)$;
    $\boldsymbol{W}_0 \leftarrow \boldsymbol{W}(1\colon \ell, 1\colon k)$;
    $\boldsymbol{\Sigma}_0 \leftarrow \boldsymbol{\Sigma}$;
  else
    $\ell \leftarrow \ell + 1$;
$\boldsymbol{A}_1 \leftarrow \frac{1}{iN}\sum_{j=1}^{N} \widehat{\Gamma}_j\boldsymbol{X}(\widehat{\Gamma}_j)^{-1}\widehat{\boldsymbol{V}}$;
$\boldsymbol{B} \leftarrow \boldsymbol{V}_0^{\mathrm{h}}\boldsymbol{A}_1\boldsymbol{W}_0\boldsymbol{\Sigma}_0^{-1}$;
$(\lambda_j)_{j\leq k} \leftarrow eigs(B)$;

At first the algorithm might seem prohibitively expensive due to the application of a singular value decomposition. However, since the number of eigenvalues k will be small and often known a-priori such that one can even choose $k = \ell$ directly. An example of this is given by accelerator cavities, which are specifically designed to have only few resonant modes close to a frequency specified by their utilisation within the accelerator. Then complexity of the singular value decomposition becomes negligible in comparison to solving the linear system in lines 5 and 15 of the algorithm.

Moreover, in an actual implementation $\boldsymbol{A}_1$ and $\boldsymbol{A}_0$ are assembled simultaneously, since the most expensive operation of the algorithm is evaluating $\boldsymbol{X}(\widehat{\Gamma}_j)^{-1}\widehat{\boldsymbol{V}}$. In general, k and ℓ will be small, so storing both $\boldsymbol{A}_j$ poses no issues.

Since smooth contours should be used, exclusively, one can expect the trapezoidal rules for the assembly of the $\boldsymbol{A}_j$ to converge exponentially with respect to N. This means only few quadrature points are needed, and the bottleneck in terms of accuracy of the entire scheme appears to be the accuracy in which $\boldsymbol{X}$ represents the bilinear form of the EFIE, cf. the left-hand side of Problem 2.37.

Remark 6.4 (*Obtaining the Eigenvectors*) Note that through this algorithm we not only obtain the eigenvalues of Problem 6.1, but also the coefficients of the corresponding eigenfunctions in the form of the matrices $\boldsymbol{V}$ and $\boldsymbol{W}$. Note moreover that the number of non-zero singular values reflects the number of solutions within D and can be used for verification of an implementation if the number of solutions is known from analytical representations or measurements.

6.2 Numerical Examples: Resonances in Perfectly Conducting Structures

In the following we will discuss some numerical experiments showcasing the application of the contour integral method to the isogeometric boundary element method. In this, we will omit a discussion of the naturally arising question of how the eigenvalues of the continuous Problem 2.32 are linked to those of the discrete counterpart Problem 6.1, and how such a link would be affected through the application of the contour integral method. These questions have been addressed by Unger [5] in detail, see also [6].

Moreover, often one will find that the number of eigenvalues k inside the contour will be $k > 1$. This means that multiple linear systems have to be solved for the same matrix. Since no quasi-optimal preconditioners for the isogeometric discretisation of the electric field integral equation are known, iterative solvers yield suboptimal runtimes. Thus, since the examples are based on comparably small systems for which the compression is still within the preasymptotic regime, the following numerical experiments will use a dense matrix assembly, cf. Sect. 5.6.

This is paired with a partially pivoted LU decomposition of [7] to solve the arising systems. As discussed in Sect. 5.5, our higher-order approaches yield systems small

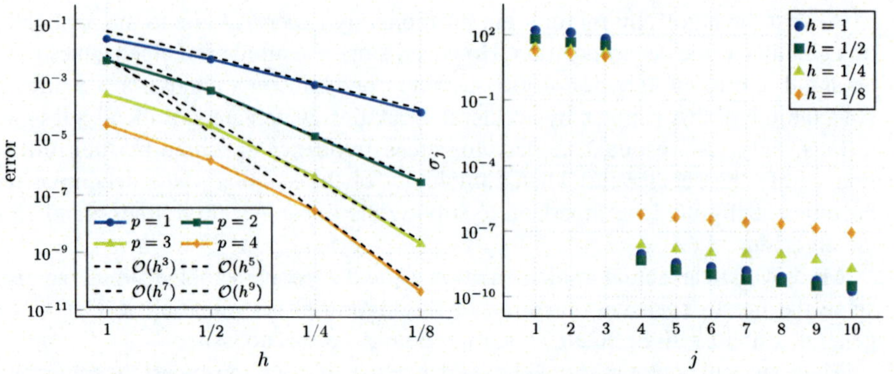

Fig. 6.1 On the left the mean absolute difference of the computed first eigenvalue of the sphere to the analytical solution is depicted. On the right the computed singular values for the example with $p = 3$ can be seen

enough for this approach to be more than feasible. This holds especially true since for $k > 1$ one has to solve multiple systems with the same matrix, and since direct solvers suffer less from ill-conditioned systems than iterative methods.

6.2.1 The Unit Sphere

As a first example, we compute the first eigenvalue of Problem 6.1 for the sphere, cf. Fig. 6.1 for the results. The contour was chosen as the curve

$$0.2 \cdot \sin(t) + i \cdot 0.2 \cdot \cos(t) + 2.7, \qquad \text{for } t = [0, 2\pi).$$

The trapezoidal rule for lines 5 and 15 of the algorithm was chosen with $N = 30$. An analytical solution λ_{ana} is known [5, Sect. 5.1.2] and given as the first root of the spherical Bessel function of the first kind, cf. Fig. 6.2.

It is an eigenvalue of multiplicity three, thus three non-zero singular values in $\mathbf{\Sigma}$ are expected. This behaviour is reflected by the numerical examples perfectly. The error in Fig. 6.1 refers to the average error of all three computed eigenvalues, i.e.,

$$\text{error} = \frac{1}{3} \sum_{0 \leq j < 3} |\lambda_j - \lambda_{\text{ana}}|.$$

The convergence behaviour w.r.t. h matches the one predicted by Corollary 3.49, compare Remark 3.50.

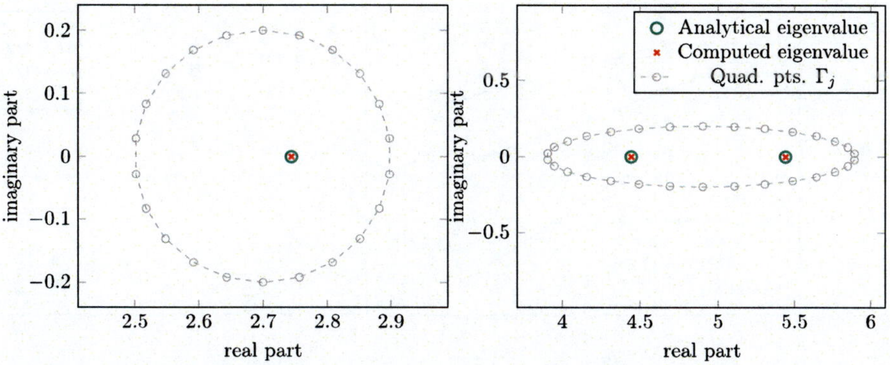

Fig. 6.2 The setup for the contour integral method. Sphere (left) and cube (right), both computed with $N = 30$, $p = 2$, $h = 1/4$

6.2.2 *The Unit Cube*

As a second example, we investigate the first three eigenvalues of the unit cube, where an analytical solution is given by the eigenvalues $\lambda_{\text{ana},0} = \pi\sqrt{2}$ of multiplicity three and $\lambda_{\text{ana},1} = \pi\sqrt{3}$ of multiplicity two. The ellipse was defined as

$$\sin(t) + i \cdot 0.2 \cdot \cos(t) + 4.9, \qquad \text{for } t = [0, 2\pi),$$

where the contour integrals were evaluated, again, with $N = 30$. The error visualised corresponds to

$$\text{error} = \frac{1}{5} \sum_{0 \le j < 5} \min_{i=0,1} |\lambda_j - \lambda_{\text{ana},i}|.$$

Again, as one can see in Fig. 6.3, the multiplicity of the eigenvalues is reflected perfectly by the non-zero singular values, i.e., all eigenvalues have been recognised. Moreover, we again observe the doubled order of convergence with respect to h.

6.2.3 *The TESLA Cavity*

As a third example we discuss an eigenvalue computation of the one-cell TESLA geometry, cf. Fig. 6.4. For this example no analytical solution is known, but experience dictates a resonant frequency around $\kappa \approx 26.5$. We choose the contour

$$2 \cdot \sin(t) + i \cdot 0.5 \cdot \cos(t) + 26.5, \qquad \text{for } t = [0, 2\pi)$$

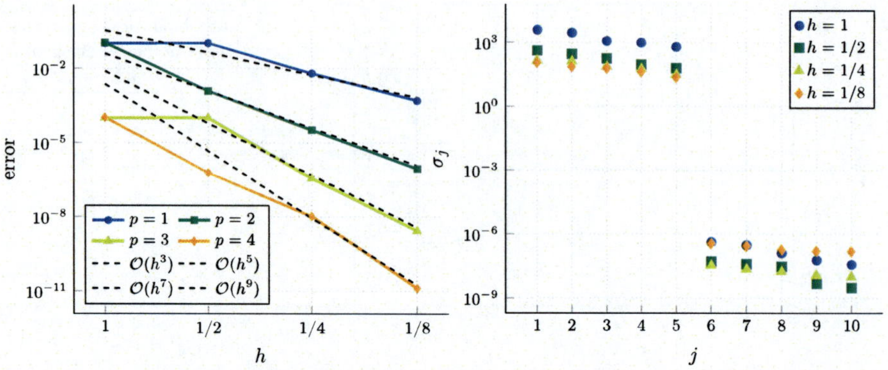

Fig. 6.3 On the left the minimal difference of the simultaneously computed first and second eigenvalue λ_{cim} of the cube to their analytical solution λ_i via $\min_{i=0,1} |\lambda_{\text{cim}} - \lambda_i|$ is depicted. On the right the computed singular values for the example with $p = 3$ can be seen

and $N = 22$. As a reference solution we utilise the result of a computation with $p = 4$ and $h = 1/8$. This reference solution was compared to a computation with CST Microwave Studio 2018. The set solver order was 3rd (constant) and the mesh was generated with 200 771 curved tetrahedral elements of order 5. CST yields the solution of 1.27666401260 GHz. Our reference solution of $\lambda_{\text{cim}} = 26.75690022$ corresponds to a frequency of 1.276664063 GHz. This results in a relative error of 3.9798e-08. Thus our experiments are in good agreement with those of commercial software. However, note that in Fig. 6.4 one can clearly see stagnation w.r.t. the CST Solution on higher-order computations and small h. This suggests that the Bembel reference solution is more accurate, provided the convergence of the contour integral approach behaves as observed in the previous numerical experiments.

The order of convergence for the cavity is not as pronounced as in the examples with analytical solution, which, in light of the results of Sect. 5.3 is most likely due to an imperfect reference solution or a more complex geometry. Either way, one can clearly see an increased accuracy in higher-order approaches.

Overall, the contour integral method yields exceptional accuracies in conjunction with our isogeometric boundary element method. Judging from the asymptotic behaviour predicted by Remark 3.50 and observed in the numerical examples, the accuracy of the combination of isogeometric boundary element method and the contour integral method promises higher orders of convergence to the correct solution compared to currently implemented volume based approaches, since these will not benefit from the convergence order of $2p + 1$, cf. Remark 3.50. This means that, for the same computational resources, the maximum reachable accuracy of the IGA-BEM with contour integration is higher compared to many volume-based methods. However, due to the long time spend solving the systems, this means that the IGA-BEM with contour integral offers this higher accuracy only in exchange for runtime.

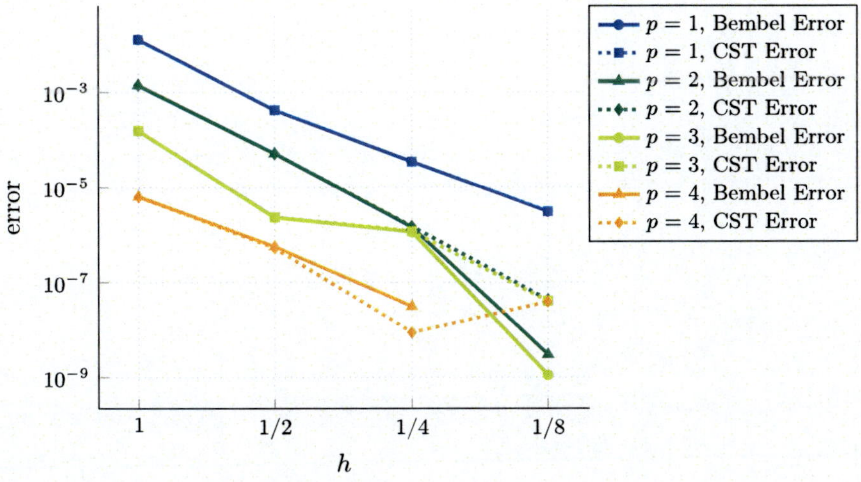

Fig. 6.4 Eigenvalue problem of the one-cell TESLA cavity solved for the first eigenvalue. Error has to be understood as the relative difference to the result of a computation. Bembel Reference refers to the error of our method to a reference computation with $h = 1/8$ and $p = 4$ of our own implementation, and the CST Reference to the error w.r.t. a reference solution obtained via CST Microwave Studio 2018

References

1. Beyn W-J (2012) An integral method for solving nonlinear eigenvalue problems. Linear Algebra Appl 436(10):3839–3863. Special Issue dedicated to Heinrich Voss's 65th birthday
2. Asakura J, Sakurai T, Tadano H, Ikegami T, Kimura K (2009) A numerical method for nonlinear eigenvalue problems using contour integrals. JSIAM Lett 1:52–55
3. Imakura A, Du L, Sakurai T (2016) Relationships among contour integral-based methods for solving generalized eigenvalue problems. Jpn J Ind Appl Math 33(3):721–750
4. Mennicken R, Möller M (2003) Non-self-adjoint boundary eigenvalue problems, 1st edn, vol 192. Mathematics Studies 1. Elsevier, Amsterdam
5. Unger G (2017) Convergence analysis of a Galerkin boundary element method for electromagnetic eigenvalue problems. Technical Report 2017/2, Institute of Computational Mathematics, Graz University of Technology
6. Kurz S, Schöps S, Unger G, Wolf F, Solving Maxwell's Eigenvalue problem via isogeometric boundary elements and a contour integral method. In preparation
7. Benoît J, Guennebaud G (2019) Eigen3 C++ linear algebra template library. Official website, https://eigen.tuxfamily.org. Accessed 20 Aug 2019

Chapter 7
Final Remarks

As closing remarks, we will summarise the results of this thesis and mention some closely related open problems.

7.1 Summary

We discussed the analysis and implementation of an isogeometric boundary element method for the electric wave equation (2.27) in three dimensions.

Chapter 3 provided definitions for all the required discrete spaces in the multi-patch setting and showed quasi-optimal approximation properties w.r.t. the de Rham complex and its traces, which were summarised in Sect. 3.5.

We introduced an isogeometric discretisation of the EFIE for which we provided a proof for the discrete inf-sup stability (Theorem 3.47). Based on this and the approximation results, we could derive quasi-optimality of the numerical solution (Theorem 3.48), as well as doubled convergence order in the pointwise error of the electric field and other continuous linear output functionals (Corollary 3.49).

We proposed an implementation based on a superspace approach (4.6) such that a modified version of the interpolation-based fast multipole method of Dölz et al. [1] is applicable, cf. Sect. 4.3.

We verified all this in a series of numerical examples in Chap. 5, also showcasing that the compression scheme maintains asymptotic behaviour for a suitable choice of parameters.

Lastly, in Chap. 6, we applied the contour integral method of Beyn [2] to the isogeometric boundary element framework to solve the underlying eigenvalue problem. Here, we provide a series of numerical examples showing that the doubled rate of

© The Author(s), under exclusive license to Springer Nature Switzerland AG 2021

F. Wolf, *Analysis and Implementation of Isogeometric Boundary Elements for Electromagnetism*, Springer Theses,
https://doi.org/10.1007/978-3-030-61939-8_7

convergence transfers to the error in the eigenvalues as well, as theoretically predicted in Remark 3.50.

These were then applied to the one-cell TESLA cavity and benchmarked successfully against commercial software. Moreover, we provide the entire codebase as open-source software [3, 4] and explain the API of the library, see [5] and Appendix B.

7.2 Things Left to Consider

While the accounts made within this thesis form a coherent storyline capable of standing on its own, there are still some closely related topics which deserve some work. We will discuss some of them.

- As mentioned before, the EFIE is notoriously ill conditioned. While isogeometric boundary element methods offer high accuracies, the time to solution depends on efficient linear solvers. Thus, when used in conjunction with compression, a quasi-optimal preconditioning strategy for iterative solvers would be beneficial. For the case for classical (lowest order triangular) elements such methods are technical to implement but understood, see e.g. the works of Andriulli et al. [7], and Christiansen and Nédélec [6] or those cited therein. However, for the isogeometric spline spaces efficient preconditioning is still an open problem.
- To resolve fine sub-structures of geometries, approaches based on tensor product refinement are unsuitable, since the refinement will propagate globally. This problem could either be overcome using mortaring techniques [8] or spline variants capable of local refinement, e.g. T-splines [9]. Also, to maintain optimal convergence behaviour, adaptive refinement strategies are required, see e.g. [10]. However, to the best of my knowledge, neither local refinement nor adaptivity have been investigated for isogeometric discretisations of the electric field integral equation.
- Concerning adaptivity, an approach via divergence-conforming wavelets could be promising. While for the lowest order case some works exist, including the required norm equivalences, see e.g. [11], higher-order wavelets and norm equivalences w.r.t. $\boldsymbol{H}_{\times}^{-1/2}(\mathrm{div}_\Gamma, \Gamma)$ have, to the best of my knowledge, not been discussed. Such an approach would be interesting since it generally yields sparse systems, thus do not need aditional compression schemes. Moreover, wavelet bases have shown to better conditioned compared to splines and classical elements when used for scalar PDEs, and they are naturally suited for adaptive refinement [12, 13].
- A last point to mention would be the handling of trimmed geometries, i.e., geometries which are not representable as NURBS parametrisation, but rather as implicitly defined subdomains thereof. While works for this exist, see e.g. [14], a clear best-practice and a scheme applicable to the electromagnetic case have yet to be developed.

References

1. Dölz J, Harbrecht H, Peters M (2016) An interpolation-based fast multipole method for higher-order boundary elements on parametric surfaces. Int J Numer Methods Eng 108(13):1705–1728
2. Beyn W-J (2012) An integral method for solving nonlinear eigenvalue problems. Linear Algebra Appl 436(10):3839–3863. Special Issue dedicated to Heinrich Voss's 65th birthday
3. Dölz J, Harbrecht H, Kurz S, Multerer M, Schöps S, Wolf F (2019) Bembel. The C++ boundary element engineering library. Official website, www.bembel.eu. Accessed 20 Aug 2019
4. Dölz J, Harbrecht H, Kurz S, Multerer M, Schops S, Wolf F (2019) Bembel. The C++ boundary element engineering library. Archived Version 0.9. Zenodo. https://doi.org/10.5281/zenodo.2671596
5. Dölz J, Harbrecht H, Kurz S, Multerer M, Schöps S, Wolf F (2020) Bembel: the fast isogeometric boundary element C++ library for laplace, Helmholtz, and electric wave equation. In: SoftwareX 11, p 10476
6. Christiansen SH, Nédélec J-C (2002) A preconditioner for the electric field integral equation based on Calderon formulas. SIAM J Numer Anal 40(3):1100–1135
7. Andriulli FP, Cools K, Bagci H, Olyslager F, Buffa A, Christiansen S, Michielssen E (2008) A multiplicative Calderon preconditioner for the electric field integral equation. IEEE Trans Antennas Propag 56(8):2398–2412
8. Buffa A, Corno J, de Falco C, Schöps S, Vázquez R (2018) Isogeometric mortar coupling for electromagnetic problems. Submitted arXiv:1901.00759
9. Bazilevs Y, Calo V, Cottrell J, Evans J, Hughes T, Lipton S, Scott M, Sederberg T (2010) Isogeometric analysis using T-splines. Comput Methods Appl Mech Eng 199(5):229–263. Computational Geometry and Analysis
10. Gantner G (2017) Optimal adaptivity for splines in finite and boundary element methods. Dissertation. TU Wien, Wien
11. Jensen TK, Christensen O, Pedersen M (2003) On adaptive wavelet-based methods for the Maxwell equations. Technical University of Denmark, Kongens Lyngby, Denmark
12. Harbrecht H (2001) Wavelet Galerkin schemes for the boundary element method in three dimensions. Dissertation. Technische Universität Chemnitz
13. Harbrecht H, Peters M (2013) Comparison of fast boundary element methods on parametric surfaces. Comput Methods Appl Mech Eng 261:39–55
14. Marussig B, Hughes TJR (2018) A review of trimming in isogeometric analysis: challenges, data exchange and simulation aspects. Arch Comput Methods Eng 25(4):1059–1127

Appendix A
Multipatch Estimates for Three Dimensions

All the presented estimates are applicable to achieve three-dimensional estimates as well. We will briefly go over the construction and state the result corresponding to Theorem 3.22.

For $p > 0$ we define the spline complex on $[0, 1]^3$ via

$$\begin{aligned}
\mathbb{S}^0_{\boldsymbol{p},\boldsymbol{\Xi}}([0,1]^3) &:= S_{p_1,p_2,p_3}(\Xi_1, \Xi_2, \Xi_3),\\
\mathbf{S}^1_{\boldsymbol{p},\boldsymbol{\Xi}}([0,1]^3) &:= S_{p_1-1,p_2,p_3}(\Xi_1', \Xi_2, \Xi_3)\times\\
&\qquad \times S_{p_1,p_2-1,p_3}(\Xi_1, \Xi_2', \Xi_3)\times\\
&\qquad\quad \times S_{p_1,p_2,p_3-1}(\Xi_1, \Xi_2, \Xi_3'),\\
\mathbf{S}^2_{\boldsymbol{p},\boldsymbol{\Xi}}([0,1]^3) &:= S_{p_1,p_2-1,p_3-1}(\Xi_1, \Xi_2', \Xi_3')\times\\
&\qquad \times S_{p_1-1,p_2,p_3-1}(\Xi_1', \Xi_2, \Xi_3')\times\\
&\qquad\quad \times S_{p_1-1,p_2-1,p_3}(\Xi_1', \Xi_2', \Xi_3),\\
\mathbb{S}^3_{\boldsymbol{p},\boldsymbol{\Xi}}([0,1]^3) &:= S_{p_1-1,p_2-1,p_3-1}(\Xi_1', \Xi_2', \Xi_3').
\end{aligned} \tag{A.1}$$

Let $f_0, \boldsymbol{f}_1, \boldsymbol{f}_2, f_3$ be sufficiently smooth. We can use the transformations

$$\begin{aligned}
\iota_0(\boldsymbol{F})(f_0) &:= f_0 \circ \boldsymbol{F}, & \iota_1(\boldsymbol{F})(\boldsymbol{f}_1) &:= (d\boldsymbol{F})^\top(\boldsymbol{f}_1 \circ \boldsymbol{F}),\\
\iota_2(\boldsymbol{F})(\boldsymbol{f}_2) &:= \det(d\boldsymbol{F})(d\boldsymbol{F})^{-1}(\boldsymbol{f}_2 \circ \boldsymbol{F}), & \iota_3(\boldsymbol{F})(f_3) &:= \det(d\boldsymbol{F})(f_3 \circ \boldsymbol{F}),
\end{aligned} \tag{A.2}$$

to define the corresponding spaces in the single patch physical domain as in (3.3), cf. [1]. Now, the projections $\tilde{\Pi}^0_{\boldsymbol{p},\boldsymbol{\Xi},\Omega}$, $\tilde{\boldsymbol{\Pi}}^1_{\boldsymbol{p},\boldsymbol{\Xi},\Omega}$, $\tilde{\boldsymbol{\Pi}}^2_{\boldsymbol{p},\boldsymbol{\Xi},\Omega}$, and $\tilde{\Pi}^3_{\boldsymbol{p},\boldsymbol{\Xi},\Omega}$ w.r.t. the reference domain for $\boldsymbol{\Xi} = [\Xi_1, \Xi_2, \Xi_3]$ defined in complete analogy to (3.9), commute with the differential operators **grad**, **curl** and div. By properties of the pullbacks, cf. [2, Sect. 5.1], this holds for the physical domain as well. The three-dimensional global B-spline projections are then defined as

© The Editor(s) (if applicable) and The Author(s), under exclusive license to Springer Nature Switzerland AG 2021

F. Wolf, *Analysis and Implementation of Isogeometric Boundary Elements for Electromagnetism*, Springer Theses,
https://doi.org/10.1007/978-3-030-61939-8

$$\tilde{\Pi}^0_\Omega := \bigoplus_{0\le j<N} \left((\iota_0(\boldsymbol{F}_j))^{-1} \circ \tilde{\Pi}^0_{p,\Xi,\Omega} \circ \iota_0(\boldsymbol{F}_j) \right), \quad \tilde{\boldsymbol{\Pi}}^1_\Omega := \bigoplus_{0\le j<N} \left((\iota_1(\boldsymbol{F}_j))^{-1} \circ \tilde{\boldsymbol{\Pi}}^1_{p,\Xi,\Omega} \circ \iota_1(\boldsymbol{F}_j) \right),$$

$$\tilde{\boldsymbol{\Pi}}^2_\Omega := \bigoplus_{0\le j<N} \left((\iota_2(\boldsymbol{F}_j))^{-1} \circ \tilde{\boldsymbol{\Pi}}^2_{p,\Xi,\Omega} \circ \iota_2(\boldsymbol{F}_j) \right), \quad \tilde{\Pi}^3_\Omega := \bigoplus_{0\le j<N} \left((\iota_3(\boldsymbol{F}_j)^{-1} \circ \tilde{\Pi}^3_{p,\Xi,\Omega} \circ \iota_3(\boldsymbol{F}_j) \right).$$

In complete analogy to the proof of Theorem 3.22, one can achieve the following result for the three-dimensional multipatch spline complex.

Corollary A.1 *Let the volumetric analogue of Assumptions 3.5 and 3.7 be satisfied. Assume the functions f_0, $\boldsymbol{f}_1$, $\boldsymbol{f}_2$, f_3 to be sufficiently smooth, i.e., such that the norms and interpolation operators below are well defined. Then one finds, for integers s as below,*

$$\begin{aligned}
\left\| f_0 - \tilde{\Pi}^0_\Omega f_0 \right\|_{H^r(\Omega)} &\lesssim h^{s-r} \| f_0 \|_{H^s_{\mathrm{pw}}(\Omega)}, && 3 \le s \le p+1, \\
\left\| \boldsymbol{f}_1 - \tilde{\boldsymbol{\Pi}}^1_\Omega \boldsymbol{f}_1 \right\|_{\boldsymbol{H}(\mathrm{curl},\Omega)} &\lesssim h^{s} \| \boldsymbol{f}_1 \|_{\boldsymbol{H}^s_{\mathrm{pw}}(\mathrm{curl},\Omega)}, && 2 < s \le p, \\
\left\| \boldsymbol{f}_2 - \tilde{\boldsymbol{\Pi}}^2_\Omega \boldsymbol{f}_2 \right\|_{\boldsymbol{H}(\mathrm{div},\Omega)} &\lesssim h^{s} \| \boldsymbol{f}_2 \|_{\boldsymbol{H}^s_{\mathrm{pw}}(\mathrm{div},\Omega)}, && 1 < s \le p, \\
\left\| f_3 - \tilde{\Pi}^3_\Omega f_3 \right\|_{L^2(\Omega)} &\lesssim h^{s} \| f_3 \|_{H^s_{\mathrm{pw}}(\Omega)}, && 0 \le s \le p,
\end{aligned}$$

for $r = 0, 1$.

References

1. Hiptmair R (2002) Finite elements in computational electromagnetism. Acta Numer 11:237–339
2. Beirão da Veiga L, Buffa A, Sangalli G, Vázquez R (2014) Mathematical analysis of variational isogeometric methods. Acta Numer 23:157–287

Appendix B
Notes on the C++ Implementation

This Appendix is devoted to briefly reviewing the structure of the code used for all computations discussed within this thesis [1]. The code, called *Bembel*, short for *Boundary Element Method Based Engineering Library,* is publicly available under the GNU GPLv3 [2].

We will not discuss all of the underlying algorithmic and software architectural decisions made, most of which are discussed within this thesis. Rather, this section will go through all of the main building blocks of the user interface and explain the underlying design considerations. This explanation will focus on the archived version 0.9 [3], which was used for the computations presented in this thesis. The current release of *Bembel* has been largely rewritten and is covered in [1].

B.1 Design Considerations

Most modern three-dimensional boundary element codes with built-in matrix compression are written in C or C++, which are proven to lead to efficient implementations. One of the central aims of Bembel is to provide a computationally efficient isogeometric boundary element code with a plain and simple user interface. Therefore, the API of Bembel is designed in modern C++ and provides an interface for the Eigen template library for numerical linear algebra [4]. This allows the user for a programming experience similar to Matlab and Octave and for the use of all matrix-free algorithms from the Eigen library. Particularly, all iterative solvers provided by Eigen can be employed without further modifications. In contrast, the low-level routines of Bembel are written in optimised C code which is tuned towards efficiency. Shared memory parallelism by the use of OpenMP is provided and hence aims at multicore machines.

© The Editor(s) (if applicable) and The Author(s), under exclusive license to Springer Nature Switzerland AG 2021

F. Wolf, *Analysis and Implementation of Isogeometric Boundary Elements for Electromagnetism*, Springer Theses,
https://doi.org/10.1007/978-3-030-61939-8

B.2 Implementation

The following sections each correspond to a central piece of the API, each implemented in one of the include/*.hpp files.

B.2.1 Geometry

With Bembel, geometry representations can directly be imported from files generated vie the octave package geopdes [5]. However, we should mention that the implementation of the Geometry class is rather flexible such that the NURBS mappings may be replaced by arbitrary smooth parametric mappings defined by the user.

B.2.2 PDE

The PDE to solve and the potential ansatz to use are defined by a derivation of the base class PDEproblem. The currently available derivations are LaplaceSingle, HelmholtzSingle, and MaxwellSingle, which implement the boundary integral equations for Laplace, Helmholtz and Maxwell problems. The derived classes contain all necessary information for the PDE and the boundary integral equation under consideration. Note that, for HelmholtzSingle and MaxwellSingle, the construction of a class instance requires a wave-number described as std::complex<**double**>.

B.2.3 Discretisation

Given a Geometry object and a class inheriting from PDEproblem, e.g. LaplaceSingle, an object of the template class Discretization<LaplaceSingle > describes a conforming boundary element space on refinement level L and polynomial degree P. The type of the B-spline boundary element space is automatically defined through the class inheriting from PDEproblem, e.g. LaplaceSingle.

B.2.4 Boundary Values

The load vectors of the Galerkin systems can be obtained via quadrature from a std:: function object describing the boundary values. For the Laplace case, computeRhs takes a Discretization and a std::function<**double**(Eigen::Vector3d)>) object and returns the load vector as an Eigen::VectorXd. For the Helmholtz and

Maxwell case, the return value is an `Eigen::VectorXcd`. Note that in this case the `std::function` object has to be modified accordingly, i.e. it must accept arguments of type `Eigen::Vector3cd`. Bembel provides exemplary functions. They can be wrapped into this framework, e.g, via C++ lambdas.

B.2.5 System Matrix

Given a specialised `Discretization`, for example `Discretization<LaplaceSingle>`, and a degree for the $\mathcal{H}^2$-compression, a corresponding specialisation of the template class `HierarchicalMatrix`, for example `HierarchicalMatrix<LaplaceSingle>`, may be used to assemble the compressed system matrix of the Galerkin system. Note that the type of the PDE and the geometry are directly encoded into the discretisation, such that no further information is required here.

The template class `HierarchicalMatrix<T>` inherits from the corresponding specialisation of Eigen's `EigenBase` matrix class, i.e. `EigenBase<HierarchicalMatrix<T>>`. Therefore, a `HierarchicalMatrix` object may use the Eigen interface and can be passed to all functions which accept `EigenBase<Derived>` objects as parameters. The Eigen methods `rows()` and `cols()` are specialised in order to return the dimensions. Moreover, to facilitate the multiplication of `Eigen::VectorX` objects, the template specialisation `Product<HierarchicalMatrix, Rhs, AliasFreeProduct>` of the Eigen `Product`class is provided. In particular, the operator `*` is overloaded to support expressions like `H * x`, where `H` is a `HierarchicalMatrix<PDEproblem>` object, and `x` an `Eigen::VectorX` object, with a fitting `Scalar` type.

B.2.6 Solution of the System of Linear Equations

The solution of linear systems within Bembel can be performed by the matrix-free solvers from Eigen. To that end, the template class `HierarchicalMatrix` inherits the `Eigen::internal::traits` of an `Eigen::SparseMatrix` with the same `Scalar` type as in the corresponding example from the Eigen documentation on matrix-free solvers. We opted for the `Eigen::SparseMatrix` traits, since they seem to impose the least requirements. Hence, in view of the implemented general matrix-vector (GEMV) product `H * x`, all Eigen iterative solvers which solely rely on this operation may be used.

B.2.7 Evaluation of the Solution

Having the density vector `rho` from the iterative solver, the solution to the PDE in either the exterior or the interior domain can be evaluated by the function `Sol::evalSolution(gridpoints, rho, myDisc)`. In here, gridpoints should be an

`Eigen::Matrix<double, Eigen::Dynamic, 3>` matrix of `n` three-dimensional points in the domain of interest. Depending on the PDE, `sol` is either an `Eigen::VectorXd`, `Eigen::VectorXcd` or `Eigen::Matrix<std::complex<double>, Eigen::Dynamic, 3>` object, where each row contains the solution at the corresponding gridpoint.

B.2.8 *Visualisation*

The `Bembel::Vis::plotDiscretizationToVTK` functions can be used to visualise `rho` and the geometry in vtk-format. All of the pictures within this thesis are based on these routines. Their functionality is explained extensively in the `src/examples/example_Visualization.cpp` file.

B.3 Example Program

In the following we discuss a full example for the solution of an electromagnetic scattering problem. It corresponds to the file `src/examples/example_MaxwellSingle.cpp`, excluding comments and the definition of the excitation.

```
int main() {
  using namespace Bembel;
  using namespace Eigen;

  const int multipoleDegree = 16;
  const int knotRepetition = 1;
  const std::complex<double> wavenumber(3, 0);

  Geometry geom("geometry.dat");
  MaxwellSingle pde(wavenumber);
  MatrixXd gridpoints = Util::makeSphereGrid(6, 10);

  const std::function<Vector3cd(Vector3d, std::complex<double>)>
      fun = ... ; // Define incident wave

  for (int P = 1 ; P < 10 ; ++P)) {
    for (int L = 0 ; L < 6 ; ++L) {
      Discretization<MaxwellSingle> disc(geom, pde, P, knotRepetition, L);
      VectorXcd rhs = Rhs::computeRhs(disc, fun);
      HierarchicalMatrix<MaxwellSingle> hmat(disc, multipoleDegree);
      GMRES<HierarchicalMatrix<MaxwellSingle>, IdentityPreconditioner> gmres;
```

```
      gmres.compute(hmat);
      VectorXcd rho = gmres.solve(rhs);
      MatrixXcd pot = Sol::evalSolution(gridpoints,
    rho, disc);
    }
  }
  return (0);
}
```

At the beginning of the program, the polynomial degree for the $\mathcal{H}^2$-compression, the knot repetition of the B-spline spaces, and the wave-number are defined. The `Geometry` object is loaded from the file `"geometry.dat"`, and the PDE is defined for a given wave-number. `Util::makeSphereGrid(6,10)` generates ten evaluation points on ten lines of latitude for the scattered field on a sphere around the origin with radius 6. Afterwards, an incident wave for the scattering problem is defined in `fun`.

The rest of the example program loops over different polynomial degrees `P` and refinement levels `L`. In the loop, the discrete space is defined as a `Discretization` object, the load vector and the system matrix are assembled and solved with the GMRES solver from the Eigen (unsupported) package. Finally, the scattered field is evaluated at the previously defined points.

Note that the program is straightforwardly adapted to the Laplace and Helmholtz cases as indicated by the example files in the `src/examples` folder. These examples also provide error measurements of the solution to the underlying PDE and output of these errors into log-files. These were not included to the above example for a better understanding of the main functionality of the code. LaTeX-files creating error plots can be found in the `LaTeX` folder. The routines can also visualise the intermediate quantity `rho`.

References

1. Dölz J, Harbrecht H, Kurz S, Multerer M, Schöps S, Wolf F (2020) Bembel: the fast isogeometric boundary element C++ library for laplace, Helmholtz, and electric wave equation. In: SoftwareX vol 11, p 10476
2. Dölz J, Harbrecht H, Kurz S, Multerer M, Schöps S, Wolf F (2019) Bembel. The C++ boundary element engineering library. Official website, www.bembel.eu. Accessed 20 Aug 2019
3. Dölz J, Harbrecht H, Kurz S, Multerer M, Schops S, Wolf F (2019) Bembel. The C++ boundary element engineering library. Archived Version 0.9. Zenodo. https://doi.org/10.5281/zenodo.2671596
4. Benoît J, Guennebaud G (2019) Eigen3 C++ linear algebra template library. Official website, https://eigen.tuxfamily.org/. Accessed 20 Aug 2019
5. de Falco C, Reali A, Vázquez R (2011) GeoPDEs: a research tool for isogeometric analysis of PDEs. Adv Eng Softw 42:1020–1034

Curriculum Vitae

Felix Wolf currently holds a position as a postdoctoral researcher at the Technical University of Darmstadt, *Institute for Accelerator Science and Electromagnetic Fields* and the *Centre for Computational Engineering*. He is a member of the working groups *Computational Electromagnetics* of Prof. Dr. Sebastian Schöps and *Electromagnetism and Mathematical Modelling* of Prof. Dr. Stefan Kurz.

He obtained his bachelor's degree in 2014 and his master's degree in 2015, both in mathematics, at the Technical University of Darmstadt.

Early in 2016, he obtained a Ph.D. scholarship of the German Excellence Initiative through the Graduate School of Computational Engineering at Technical University of Darmstadt. There, he began working on the analysis and implementation of isogeometric boundary element methods under the supervision of Prof. Dr. Stefan Kurz and Prof. Dr. Sebastian Schöps. In 2017, he obtained a full researcher position funded by DFG grants KU1553/4-1 and SCHO1562/3-1.

This book presents the results of his work on electromagnetic problems, for which he was awarded the degree of *Doktor-Ingenieur (Dr.-Ing.)*, in December 2019. In 2018, he was awarded the *Young Scientist Award* and *Best Paper Award* of the German branch of the International Union of Radio Science (U.R.S.I).

© The Editor(s) (if applicable) and The Author(s), under exclusive license to Springer Nature Switzerland AG 2021

F. Wolf, *Analysis and Implementation of Isogeometric Boundary Elements for Electromagnetism*, Springer Theses,
https://doi.org/10.1007/978-3-030-61939-8

Index

© The Editor(s) (if applicable) and The Author(s), under exclusive license
to Springer Nature Switzerland AG 2021
F. Wolf, *Analysis and Implementation of Isogeometric Boundary Elements for Electromagnetism*, Springer Theses,
https://doi.org/10.1007/978-3-030-61939-8